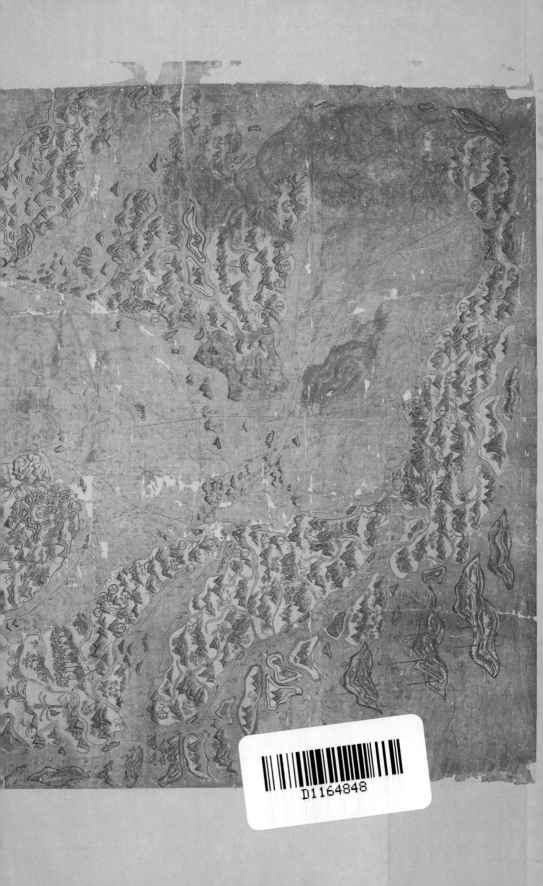

Mr. Selden's Map of China

ALSO BY TIMOTHY BROOK

The Troubled Empire: China in the Yuan and Ming Dynasties (2010)

*Vermeer's Hat: The Seventeenth Century and
the Dawn of the Global World* (2008)

Death by a Thousand Cuts
(2008; with Jérôme Bourgon and Gregory Blue)

Collaboration: Japanese Agents and Chinese Elites in Wartime China
(2005)

The Chinese State in Ming Society (2005)

Opium Regimes: China, Britain, and Japan, 1839–1952
(2000; with Bob Wakabayashi)

Documents on the Rape of Nanking (1999)

The Confusions of Pleasure: Commerce and Culture in Ming China
(1998)

*Praying for Power: Buddhism and
the Formation of Gentry Society in Late-Ming China* (1993)

*Quelling the People: The Military Suppression
of the Beijing Democracy Movement* (1992)

Mr. Selden's Map of China

Decoding the Secrets of a Vanished Cartographer

Timothy Brook

BLOOMSBURY PRESS
NEW YORK · LONDON · NEW DELHI · SYDNEY

Published by Bloomsbury Press, New York

All papers used by Bloomsbury Press are natural, recyclable products made
from wood grown in well-managed forests. The manufacturing processes
conform to the environmental regulations of the country of origin.

LIBRARY OF CONGRESS CATALOGING-IN-PUBLICATION DATA HAS BEEN APPLIED FOR

ISBN: 978-1-62040-143-9

First published in the U.K. by Profile Books Ltd in 2013
First U.S. Edition 2013

1 3 5 7 9 10 8 6 4 2

Typeset in Bulmer by MacGuru Ltd
Printed and bound in the U.S.A. by Thomson-Shore Inc., Dexter, Michigan

This book is for my uncommon reader,
Fay Sims.

It is also dedicated to the memory of
Neil Burton, fellow traveller always one
step ahead.

Item. I give and bequeathe to the said Chancelor Masters and Schollars a Mapp of China made there fairly and done in colloure together with a Sea Compasse of their making and Devisione taken both by an englishe comander who being pressed exceedingly to restore it at good ransome would not parte with it.

codicil to John Selden's will
11 June 1653

Contents

Illustrations   xi
Dramatis Personae   xiii
Dramatis Loci   xv
Timeline   xvii
Preface   xix

1. What's Wrong with this Map?   1
2. Closing the Sea   19
3. Reading Chinese in Oxford   45
4. John Saris and the China Captain   67
5. The Compass Rose   87
6. Sailing from China   110
7. Heaven is Round, Earth is Square   129
8. Secrets of the Selden Map   149
    Epilogue: Resting Places   175

Appendix I. Boxing the Chinese Compass   183
Appendix II. Coast Comparison   186
Acknowledgements and Sources   187
Index   201

Illustrations

1. The Selden map (Bodleian Library)
2. Extract from the codicil of John Selden's will, dated 11 June 1653 (Public Records Office, Kew)
3. Portrait of John Selden as a young man (© National Portrait Gallery, London)
4. Portrait of Ben Jonson (© National Portrait Gallery, London)
5. Portrait of Huig de Groot (Grotius) as a young man (courtesy of Stichling Museum, Rotterdam)
6. Studio of Peter Lely, portrait of John Selden (© National Portrait Gallery, London)
7. Portrait of Thomas Hyde (Bodleian Library)
8. Charles I touching for the King's Evil (© Royal College of Physicians)
9. Godfrey Kneller, portrait of Michael Shen, also known as *The Chinese Convert*, 1687 (© Royal Collection Trust)
10. The compass rose on the Selden map
11. The Laud rutter, *Shunfeng xiangsong* ('Dispatched on Following Winds'), title romanised by Michael Shen and translated into Latin by Thomas Hyde (Bodleian Library)
12. John Selden's Chinese compass (Bodleian Library)
13. The routes on the Selden map
14. Map of England and part of Scotland illustrating the coastline under the jurisdiction of the King's Chambers, in Selden, *Of the Dominion, or, Ownership, of the Sea* (1652), p. 366
15. Map of Great Britain and its surrounding seas, in Selden, *Of the Dominion, or, Ownership, of the Sea* (1652), p. 185
16. The starting point of the routes on the Selden map: Zhangzhou, Quanzhou and the Taiwan Strait
17. The Calicut cartouche on the Selden map
18. Samuel Purchas as he appears on the title page of his *Purchas his Pilgrimes*, 1625 (© National Portrait Gallery, London)

19. The Hondius map, in Samuel Purchas, *Purchas his Pilgrimes* (1625), vol. III, p. 360 (British Library: The British Library Board)
20. The Saris map, in Samuel Purchas, *Purchas his Pilgrimes* (1625), vol. III, p. 401 (British Library: The British Library Board)
21. *A Complete Map of Mountains and the Seas of the Earth* (*Yudi shanhai quantu*)
22. John Speed, *Asia with the Islands Adjoining*, 1626, courtesy of Taiwan Commercial Press
23. *General Topographical Map by Province of the Divisions and Correspondences of the Twenty-Eight Lunar Mansions of the Ming Dynasty*, in Yu Xiangdou, *Wanyong zhengzong* ('Complete Source for a Myriad Practical Uses') (1599), 2.2b–3a
24. 'Heaven is round, earth is square', in Zhang Huang, *Tushu Bian* ('Documentarium') (1613), 28.2a
25. Map of China, in Luo Hongxian, *Guang yutu* ('Enlarged Terrestrial Atlas') (1555)
26. The Selden map geo-referenced
27. Prince Giolo, 1692
28. T. Murray, portrait of William Dampier, 1698 (© National Portrait Gallery, London)

Dramatis Personae

Will ADAMS (1564–1620): English pilot shipwrecked in Japan in 1600 while serving on a Dutch ship; captained several voyages for the East India Company between Japan and South-East Asia, 1614–18

Richard COCKS (1566–1624): English merchant and head of the trading post established by the East India Company in Japan between 1613 and 1623

GIOLO [*geeolo*] (*c.* 1661–1692): Pacific Islander captured by Muslim slave traders in the 1680s and sold into service in Mindanao; died in Oxford in 1692

Thomas HYDE (1636–1703): Oriental scholar, appointed Assistant Keeper of the Bodleian Library in 1659 and Keeper in 1665, a post he held until 1701; appointed Laudian Professor of Arabic in 1691 and Regius Professor of Hebrew in 1697; annotator of the Selden map

Ben JONSON (1572–1637): poet, satirist, playwright, entertainer at the court of King James I, bosom friend and admirer of John Selden

William LAUD (1573–1645): appointed Bishop of London in 1628, elected Chancellor of Oxford in 1630, consecrated Archbishop of Canterbury in 1633; executed by Parliament in 1645

LI Dan [*lee dan*] (b. 1560s; d. 1625): 'China Captain' of Japan, or head of the Chinese community in Hirado; landlord of the East India Company factory; business associate of Richard Cocks; mentor of Zheng Zhilong, whose son Zheng Chenggong founded the Eastern Calm kingdom on Taiwan

Samuel PURCHAS (before 1577–1626): chaplain turned editor who published a series of popular collections of travellers' tales, starting in 1613 with *Purchas his Pilgrimage* and culminating in 1625 with *Purchas his Pilgrimes*; erstwhile friend of John Selden and acquaintance of John Saris

John SARIS (1579/80–1643): employee of the East India Company in Bantam, 1605–9; commander of the Company's Eighth Voyage, 1611–14

John SELDEN (1584–1654): lawyer, Orientalist, legal historian, parliamentarian, constitutional theorist, author of *The Closed Sea*

Michael SHEN Fuzong (*c.* 1658–1691): son of a Nanjing doctor and disciple of Jesuit missionary Philippe Couplet; sojourned in Europe between 1683 and 1691; annotator with Thomas Hyde of the Selden map

John SPEED (1542–1629): engraver, cartographer, historian of England; publisher of England's first world atlas in 1627

ZHANG Huang [*jang hwong*] (1527–1608): native of Jiangxi province, failed examination candidate, head of the prestigious White Deer Grotto Academy; compiler of the massive encyclopaedia *Tushu bian* ('Documentarium')

ZHANG Xie [*jang syeh*] (1574–1640): native of Zhangzhou, denizen of Moon Harbour, graduate of the 1594 Fujian provincial examination; author of *Dong xi yang kao* ('Study of the Eastern and Western Seas')

Dramatis Loci

BANTAM, also Bantan, Bantem: a city-state at the western end of Java; the first trading port for Europeans arriving in the South China Sea and home for John Saris, 1604–9; eclipsed by Batavia after 1619

BATAVIA (Jakarta): a port city in west Java, occupied by the Dutch in 1619 and made the base of operations of the Dutch East India Company (VOC)

HAINAN ISLAND: large island off the south coast of China's Guangdong province, known as Qiongzhou prefecture in the Ming dynasty

HIRADO: port town in Kyushu, near Nagasaki; as of 1609, one of the few ports in Japan where Chinese and European traders were permitted to reside; home for a time of Li Dan and Richard Cocks

PARACEL ISLANDS, also Western Shoals (Xisha), also Hoàng Sa Islands: a scattering of tiny islands in the north-western quarter of the South China Sea claimed by China and Vietnam

RYUKYU ISLANDS: a string of islands, of which the largest is Okinawa, between Japan and Taiwan; an independent kingdom that submitted tribute to Ming China but was under Japanese domination from the sixteenth century; formally annexed to Japan in 1895

SPRATLY ISLANDS, also Southern Shoals (Nansha): a scattering of tiny islands north-west of Borneo in the South China Sea, claimed by China, Vietnam, Brunei, Malaysia, Taiwan and the Philippines

THE TEMPLE: area of London between Fleet Street and the Thames, former home of the Knights Templar and latterly the precinct of the Inner and Middle Temples, two of the four Inns of Court to which English barristers are affiliated; site of John Selden's office in the Inner Temple and of his grave in the Temple Church

TERNATE: a small island in the Moluccas (Malukus), or Spice
   Islands, centre of the spice trade in the seventeenth century; first
   'discovered' by Portuguese in 1512; visited by Francis Drake in 1580
   and John Saris in 1613; co-occupied by Spain and the Netherlands
   from 1607 to 1663

Timeline

1600 East India Company (EIC) founded in London
Will Adams shipwrecked on the coast of Japan

1602 Dutch East India Company (VOC) founded in Amsterdam
Thomas Bodley opens the Bodleian Library in Oxford
John Selden leaves the University of Oxford for the Inns of Court
in London

1603 Jacob van Heemskerck seizes Portuguese vessel *Santa Catarina*
at Johor

1604 John Saris arrives in Bantam as an employee of the EIC

1607 the VOC sets up a base on Ternate

1608 Jodocus Hondius in Amsterdam publishes his atlas, *Map of the
World*

1609 Zhang Huang dies
Huig de Groot publishes *Mare Liberum* ('The Free Sea')
John Saris returns to London from Bantam
Tokugawa shogunate annexes Ryukyu

1611 John Saris leaves London in command of the EIC's eighth
voyage to Asia

1612 William Shakespeare stages *The Tempest* for James I
John Selden is called to the Bar

1613 John Saris reaches Japan, appoints Richard Cocks as EIC factor
at Hirado
Samuel Purchas publishes *Purchas his Pilgrimage*, dedicated by
John Selden
Zhang Huang's encyclopaedia *Tushu bian* ('Documentarium')
published

1614 Ming court slashes funding for naval patrols along the coast
John Saris returns to England

1617 Zhang Xie completes his *Study of the Eastern and Western Seas*

1618 James I questions John Selden on his *The Historie of Tithes*

1619 VOC seizes control of Jakarta, renames it Batavia

1620 the Wanli emperor dies

Ben Jonson stages *News from the New World Discovered in the Moon* for James I

1621 John Selden detained by James I for 'reasons of State knowne unto himself'

1624 Richard Cocks dies on his return voyage to England

Li Dan's trading network collapses

1625 Samuel Purchas publishes *Purchas his Pilgrimes* with two maps of China

1627 John Speed publishes his atlas, *Prospect of the World*

1628 Ming China reimposes a ban on ocean-going ships

1629 John Selden imprisoned by order of Charles I

1630 William Laud elected Chancellor of the University of Oxford

1635 John Selden publishes *Mare Clausum* ('*The Closed Sea*')

1644 the Manchus invade Ming China and absorb it into their Qing dynasty

1652 Marchamont Nedham publishes an unauthorised translation of *The Closed Sea*

1654 John Selden dies in London

1659 Selden's library delivered to the Bodleian Library

Thomas Hyde appointed Assistant Keeper of the Bodleian Library

1661 Zheng Chenggong drives VOC from Taiwan, founds Eastern Calm kingdom

1665 Thomas Hyde appointed Keeper of the Bodleian Library

1683 Qing dynasty destroys Eastern Calm kingdom, annexes Taiwan

Elias Ashmole opens the Ashmolean Museum in Oxford

1687 Michael Shen visits Thomas Hyde in Oxford to catalogue the Chinese books

1691 Michael Shen dies off Mozambique en route back to China

1692 Giolo arrives in London, dies in Oxford

Preface

Rarely does an old map make front-page news, but the map of the world that Martin Waldseemüller produced in 1507 did just that when the Library of Congress acquired it in 2003. The Waldseemüller map has been called America's birth certificate, and it cost the nation $10 million. It is beautiful, certainly, printed from twelve woodblocks so finely carved that the Jesuit schoolteacher who rediscovered the map in 1901, Joseph Fischer, assumed it to be the handiwork of the great artist Albrecht Dürer. It wasn't, but it was worthy of the mistake. As many as a thousand copies of this enormous map of the world may have been printed from these woodblocks, yet the only copy to survive is the one now on display in the foyer of the Library of Congress.

The map fetched the price it did because of one tiny detail. This is the first map on which the name America appears. Martin Waldseemüller inscribed it on a blank space in South America, roughly where we would locate Paraguay. Quite how much of the wraith-like landform snaking its way up the left-hand side of the map from the Antarctic to the Arctic the term was meant to name is unclear, but the Congress of the United States agreed that it covered enough to satisfy them. So there it is: a new name for a new continent, and all because Waldseemüller was

a big fan of the explorer–geographer Amerigo Vespucci. Had he been an enthusiast of Christopher Columbus, he might have called the new continent Columbia. But no, for him Vespucci was the discoverer of the New World.

Nine years after the map was published, Waldseemüller abandoned his innovative model of the world for a very different design, thereby rendering the 1507 original redundant. It was now a map without a future. This one copy survived only because a free-spirited priest-turned-mathematician named Johannes Schöner bought and preserved it some time before he died in 1547. He put it in a leather-bound portfolio, which ended up in Wolfegg Castle in southern Germany. It came to light only because in 1901 the castle archivist, Hermann Hafner, heard that a schoolteacher just across the border in Austria was interested in historical documents and offered him the run of the castle library. That schoolteacher, Joseph Fischer, was a Viking enthusiast looking for sources on the early Norse voyages. Without all these serendipitous connections, the map might never have crossed the five centuries that separate us from Waldseemüller. Johannes Schöner, the actor in this history closest to its beginning, feared the indifference with which objects by which one can investigate the past – indeed anything – could be treated. 'You know the times', he complained in 1533. The arts and sciences 'are so silent and neglected, it may be feared that the idiots will wipe them out'.

The book you are about to read revolves around a different map, the Selden map, so called because an English lawyer by the name of John Selden bequeathed it to the Bodleian Library in Oxford in 1654. The most important Chinese map of the last seven centuries, it maps the slice of the world that Chinese at the time knew, from the Indian Ocean in the west to the Spice Islands in the east, and from Java in the south to Japan in the north. It exists today because it came into the hands of John Selden, who shared Johannes Schöner's passion to ensure the survival of knowledge, and not just English knowledge but all knowledge, even Chinese, although it was a language he couldn't read. It is fortunate that he did so, for unlike the thousand Waldseemüllers that were printed, the Selden map is a singleton, drawn and painted by hand, the only one of its kind.

It is a large map, measuring 160 cm (63 in.) in length and 96½ cm (38 in.) in width. That makes it only half the size of the Waldseemüller (16⅔ sq. ft compared to 34 sq. ft), but still it must count as the largest wall map of its time and place. As neither China nor Europe produced sheets of paper that large, making wall maps on this scale required ingenuity. The largest sheet of paper available to the man who drew the Selden map was 65 × 128 cm (25½ × 50½ in.). He solved the size problem by taking two sheets, cutting one lengthwise down the middle and gluing one of the halves down the side of the other sheet, then trimming the length of the remaining half and gluing it along the bottom. Waldseemüller worked with smaller sheets of paper (42 × 77 cm, 16½ × 30½ in.). Rather than glue them together, he divided his map into twelve sections, printed it on twelve sheets from twelve separate woodblocks and left it to the buyer to assemble them into a single map. Then map design changed, and all the buyers but one threw their dozen sheets away. Schöner's set survived only because it disappeared into a library, which is just what happened to the Selden map. Both have now re-emerged – Waldseemüller's a century ago, Selden's just a few years back – to great public interest.

Both maps are terrifically important, in different ways. Waldseemüller drew his map just at the moment when the New World was coming into view. Europe's novel encounter with the world forced him to bend the existing mapmaking template to breaking point, and then to abandon it nine years later in favour of a new geometry better capable of encompassing the entire globe. So too in its way the Selden map bore the impact of China's encounter with the same world, seen from the other side of the globe. The man who drew the map acknowledged long-established traditions of how to draw China, but he also stepped outside that tradition to picture the lands that lay beyond China in a fashion no other Chinese cartographer had ever done. Not unlike Waldseemüller, he re-designed the world in response to an avalanche of new data about how the lands and seas beyond his native place actually lay on the surface of the earth. He also created a thing of considerable if subtle beauty, wallpapering the land mass of eastern Asia with mountains, trees and flowering plants – and the occasional whimsical detail. The two errant butterflies fluttering about in the Gobi Desert are my favourites.

It took a century for the map that names America to find its new home in the Library of Congress, where it occupies what many regard as its rightful place in the pantheon of foundational documents celebrating their nation. Will fate touch the Selden map in the same way? Painstakingly (and expensively) restored in 2011, it is now on display in the Bodleian Library. Will its story end there? Should some decide that this map has a foundational role to play in the celebration of China's national identity, its future could become complicated. But the Selden map is not China's birth certificate. Neither the Chinese name for China – Zhongguo – nor the name of the reigning dynasty – Ming – appears on it, but then China has been around for so long that neither would carry significant weight at this late moment in its history.

Not a birth certificate, then, but potentially an adoption certificate? China is currently in dispute with every maritime nation in East Asia over who may rightfully claim sovereignty over the thousands of islands that dot the East and South China Seas. The best-known, because most noisily contested, are the Diaoyu Islands north-east of Taiwan, and the Paracel and Spratly Islands in the South China Sea. As the Selden map is the only detailed and geographically specific Chinese depiction of these waters before the nineteenth century, some hope that this long-lost map may be the winning card in the diplomatic game China plays with its neighbours. Over the course of this book I will indicate my doubt in this regard and show that the Selden map has nothing to say about such topics. But patriotic sentiment and national interest are powerful forces against knowledge for its own sake, so who can say? The Selden map has been valued for insurance purposes at three-fifths the price of the Waldseemüller map. This is an arbitrary estimate for an object that has been off the market for almost four centuries. If it ever goes back on, the bidding will surely go much higher.

I have not devoted an entire book to a single map in order to deliver an *Antiques Roadshow* punchline. Rather, I take the map as an occasion to explore the age in which it was made. It was an age of remarkable creativity and change. New vistas were opening, old horizons faltering, accepted truths giving way to controversial new ideas. Ordinary people in their hundreds of thousands were on the move in search of work, survival and adventure. Ships in their tens of thousands were sailing from every port in Europe and Asia. Commodities produced on one continent were reshaping economies on another. Against this background William

Shakespeare was premiering *The Tempest*, Ben Jonson inventing the musical to amuse King James I, and John Donne being pressured by that same monarch to give up love poetry for sermon-writing, and excelling at both. John Selden was among this crowd, living life to the full in London and dutifully churning out poems while he was supposed to be studying law. The poems were decidedly second-rate: the younger man had yet to find his metier. His monumental achievements in Oriental scholarship and constitutional law lay ahead of him. But he too would change the fabric of English society just as surely as these more famous authors did. And as all this unfolded, the map that bears his name would come into his hands.

I do not begin the book with the map itself, for there are many other things to think about before we ever get to the Selden map. We have to dig first in other fields, in part because there exists basically no documentation that can tell us anything about the map. The map itself complicates its own story by having travelled half-way around the world and ending up among people who viewed it very differently from the man who made it, thereby doubling the stories that can be told about it. Far more than just a passive illustration of its age, it is a densely worked document that will reveal much about the times and places in which it was drawn, viewed and graffitied. Knowing both less and more than what the mapmaker knew, we will have to do much digging to find out how to read it.

Odd as this may seem, one book is not enough to open all the doors hidden in the details of the map, let alone travel all the corridors that lead from these doors, still less to enter all the rooms that open off the corridors. Those I have been able to enter have disclosed a mad variety of events and personalities that I never expected to encounter when I first looked at the map. They include the burning of Japanese erotica in London, the trade policies of Emperor Wanli, the design of the Chinese compass, Samuel Taylor Coleridge's intentional misspelling of Xanadu, the donation of human remains to the Bodleian Library, and the ancestral church of the Knights Templar, to mention but a few. The only topic among these that I could have predicted was the compass; everything else came as a surprise. But all of it must be taken into account if we hope to give the Selden map, about which nothing certain is known, the history it deserves.

In the end, this book is not really about a map. It is about the people whose stories intersected with it. The venture succeeds if I can demonstrate how rich, how complicated and how globally networked this era was. The map stands as a reminder – a warning, even – that our understanding of our own time will be enfeebled if we remain ignorant of the earlier practices of gaining wealth and power that have led us to our present impasse. Of course, no one back in the seventeenth century could have anticipated that the small-scale deals and conflicts going on around the South China Sea were early rumblings of the age of empire to follow, or of the age of state-corporate alliances in which we find ourselves. The traders and sailors travelling across the surface of the Selden map were simply in it for the money and thought no more about it. Curious that a desire so uninteresting could remake the world. But then why should we presume to think that our age is any different from theirs? As Johannes Schöner so bluntly put it, 'You know the times.'

1
What's Wrong with this Map?

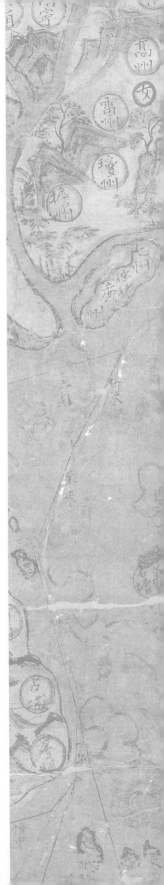

In the summer of 1976 I left China through Friendship Pass. As the train crawled into northern Vietnam through the rounded hills south of the pass, we gazed down into steep gullies crisscrossing the landscape beneath us. In some of the deeper gullies narrow streams gurgling with spring water were left to follow their natural courses. In others, the streambeds had been widened into rice paddies, the heads of the rice plants still green and not ready to harvest. An overturned steam locomotive lay in one of those gullies, its charred carcass sprawled on its back like some ruined Jurassic beast. Signs of the Vietnam War, which had ended just a year before, still littered the landscape, occasionally dramatically – beneath every railway bridge lay the twisted girders of all the other bridges it had replaced – more often inconspicuously. Already the war was being forgotten. The very landscape seemed ready to forget it. Looking down on the locomotive, I could imagine the subtropical vegetation of the gully simply growing up around the defeated machine and gently swallowing it from sight before the recovery crews could arrive.

Friendship Pass is the Orwellian name for the rail junction connecting big-brother China to little-brother Vietnam. Honoured as a site of friendship between the two countries, it has just as often been a barrier of animosity across which the two sides have eyed each other suspiciously, and occasionally launched a wasteful invasion. It would be China's turn to invade in 1979, but that piece of folly was still three years off that peaceful and beautiful summer when I came through the pass. I was leaving China at the end of a two-year stint as an exchange student, heading home via a long detour that would take me through Laos, Burma, India and Afghanistan.

We approached the pass from the north. The Chinese train shuddered to a halt, and everyone had to alight to go through border inspection inside the station before switching to the Vietnamese train, which ran on a narrower gauge. Those who weren't Chinese or Vietnamese – there were only two of us – were set aside for special treatment. When my turn came, the brusque customs officer asked me to open my backpack so that he could inspect the contents. He was looking for something, and in no time he found it.

A month before leaving Shanghai, I had gone to the customs office to arrange the shipping of my books and few possessions back to Canada prior to my departure by train through Vietnam. I had to unpack and present everything for the inspection of the customs official whose job it was to check what foreigners were sending out of China. The official, a man at mid-career wearing the uniform of the customs department, was pleasant enough; he was also thorough. After going through my books and papers closely, he set aside two things I could not send out of the country. Both were maps. One was a national atlas, the other a wall map of China. I had bought both at the Nanjing Road branch of New China Bookstore, the official – virtually the only – book retailer in the country, and still had the receipts to prove it. They were not marked 'for internal circulation', the label printed inside the vast majority of books, which we, as foreigners, were forbidden from buying. We had access only to 'open circulation' publications. It was one of those amusing Möbius strips of Cultural Revolution reasoning: the dignity of the nation would not permit Chinese to know everything foreigners knew, but it would not permit foreigners inside China to know the portion of what we knew that Chinese knew.

When I pestered the customs official in Shanghai to know why I couldn't keep them, he blandly pointed out that of course I could keep them; I just couldn't send them out of the country. When I pushed a little harder, he closed the subject down by informing me that maps had a bearing on national security. In those days, and probably these days as well, national security was the ultimate trump card of Chinese officials seeking to restrict foreign students' access to Chinese society. What that bearing actually was, no one could say. The only maps I was permitted to keep were the approved tourist maps of those cities that were open to tourists. These representations deliberately distorted space, on the flawed understanding that, should an enemy air force seek to bomb the country, these maps would confuse the pilots and cause them to miss their targets. (I know this sounds ridiculous, but those were ridiculous times.) I took the atlas and map back to my dormitory room, pondering what to do with them. The atlas was a hardback too cumbersome to consider carrying in my backpack across the length of Asia, so I gave it to a Chinese friend, who was happy to have it.

The map was another matter. I didn't want to get rid of it. It was light and could be folded into a compact square. Why not just carry it out in my backpack? Besides, the customs inspection had piqued my interest. I unfolded the map and looked again. What would have bothered the customs official? What was wrong with this map?

Nothing, as far as I could tell at first glance. Gradually it dawned on me that the map's liability had to do less with anything inside China than with its edges, the places where China abutted its numerous neighbours. I knew that China had exchanged fire with both the Soviet Union and India over disputed borders; there may have been others as well. Was this map claiming more territory for China than it had the right to occupy? Then I glanced at the South China Sea. This large and relatively shallow body of water south of China is bounded on its other three sides by Vietnam, Malaysian Borneo, Brunei and the Philippines. China has declared ownership of the whole thing, minus the standard 12 nautical miles (13.8 miles or 22.2 kilometres) that international law permits every coastal nation to claim. This is China's most egregious unilateral claim over a frontier. There it was on my map, marked out as a series of nine dashes dipping down from the main body of the country to enclose all of the South China Sea.

The Vietnamese know this same body of water as the East Sea. Their particular interest is in the scattering of some thirty-odd tiny islands known to Europeans as the Paracels. (The Portuguese brought the word from southern Brazil, where it was the native term for a protective offshore reef.) Vietnamese call them the Hoàng Sa Islands, the Yellow Shoals; to Chinese they are the Xisha, the Western Shoals, so named because there is another, wider scattering of islands 700 nautical miles to the south-east, off the north-west coast of Borneo. These they call the Nansha or Southern Shoals. The rest of us know them as the Spratly Islands (named after the English captain Richard Spratly, who sailed by them in 1843 and published an account of his voyage in London). The tiny islands clustered in these zones number in the thousands, depending on how many outcroppings that disappear at high tide deserve to be dignified with the name of island. Historically uninhabited, they are uninhabitable without sustained support from elsewhere.

China's claim has put it in a decades-long stand-off with all its South-East Asian neighbours. The first serious conflict blew up in January 1974, when China and South Vietnam fought the two-day Battle of the Paracel Islands. (Vietnam lost.) It was a useful propagandistic distraction for both sides. Unfortunately for us exchange students in China at the time, this little exercise in force majeure resulted in a lot of bombastic patriotic poetry, which our teachers forced us to read.

And there I was in the summer of 1976, heading for Vietnam with a restricted national map of China in my backpack. The stern-faced border guard saw me coming. He opened the backpack, looked through its contents and pulled out the map. Barely concealing his pleasure at being a cog in the machinery of state, he asked me why I was taking the map out of China when I had already been explicitly warned not to do so. The penny dropped. He knew all about the customs inspection in Shanghai. The incident was in my official security dossier – a fat file that it would amuse me to read some day, although that day will never come – and he had read it. This was a surprise. Remember that this was 1976. Security files had not yet been computerised; photocopiers were rare; and anyway, the first rule of security officers is never to let information out of their control. But I was a foreigner, and by simple logic a high security risk. Wherever I went, my file followed me like a patient dog. My exit permit said I was leaving through Friendship Pass, so that is where my file would wait for me.

I had no answer that would do anything but incriminate me further to his satisfaction. I could hardly try explaining to him that China's hypersensitivity about maps was based on a fetish, not on reality. The map was only as real as the paper it was printed on, a transient representation that could be altered or denied at will. To me it was merely a useful object, something it would be hard to replace back home, and certainly not for the price I had paid for it. The border guard started from another point entirely. A map did not merely represent China's sovereignty: it *was* that sovereignty. For him, the map existed at a level of reality higher than the real world. The paper was less real than the nation itself.

Map fetish is hardly unique to China. We all invest objects with a significance that, without us, they would not have. In the days of monarchy the body of the sovereign was treated as a fetish, the physical embodiment of the sacred, and anyone who transgressed it was guilty of high treason. Now that we are past the age of monarchies, this primitive aura has been sublimated and transferred to the body of the nation. A king could lose a bit of territory – and frequently did when he had to marry off one of his children – and no one would call it a sacrilege. But stick him with a penknife, and the highest possible treason had been committed. Modern states are immune to penknives, but not to a neighbour that would claim the tiniest bit of territory. Take an inch and its entire legitimacy is threatened. Move a boundary on a map and the same terrible indignity followed. So long as the national map stands in for the sacred nation, more real in some ways than the nation itself, a regime anxious about its legitimacy cannot afford to let it out of its sight. I had no choice but to leave the map at the border post before continuing south into a less desperate version of state socialism.

Twenty-five years later China once again asserted its sovereignty, this time with much higher stakes. On 1 April 2001 two Chinese Navy Finback jets intercepted a US Navy aircraft flying off the coast of China in the direction of Okinawa. According to the United States, Mission PR32 was 'a routine reconnaissance mission'. This particular plane, an Aries II with four propellers, had crossed this stretch of ocean before and was on a direct route to base when the interception occurred. In the cat-and-mouse world of aerial reconnaissance the incident was standard fare. Nations routinely scramble jets whenever they detect a foreign military plane inside their airspace. The two pilots sent out that morning had buzzed US planes before, nor was this a new experience for the twenty-three crew members on board the Aries. This sort of thing goes on all the time.

The two single-seater Finbacks closed in on the rear of the Aries with the intention of getting as close as possible, playing chicken, then turning tail and rocking it with jet wash. The pilot of the lead jet, Lieutenant-Commander Wang Wei, had experience in intercepting American planes before. So too the US commander, Lieutenant Shane Osborn, was a veteran of aerial manoeuvres over the South China Sea. Both men knew what they were doing. Both knew the rules of this particular game.

Wang brought his Finback up alongside the slower Aries, cruising off the left wing of the American plane at a distance of as little as ten feet. This is a dangerous move at 180 knots: you have to be an excellent pilot to manoeuvre in that close to another aircraft at that speed. After sitting off the left wing for a short while, Wang peeled away from the prop plane, circled and came up again from behind. What happened next depends on whom you believe. The pilot in the second Chinese jet later accused Osborn of 'veering at a wide angle' towards Wang's aircraft and ramming it. Osborn insisted that he stuck to regular procedure, which was to keep the Aries on a steady bearing for Okinawa and not alter his course. The problem was that Wang came up too fast on his second pass. When he tried to cut his speed by pulling up just as he reached the Aries's wing, he misjudged either his speed or his distance. The jet pitched upwards and caught the blades of the propeller of the outside engine on the Aries. The propeller sliced the Finback in half. Its nose spun forward and collided with the front of the US plane, while the cockpit and fuselage rammed the underside of the Aries before rocketing sideways under the right wing, narrowly missing the propellers. Some of the American crew believed they saw Wang eject, but no trace of him was ever found.

The Aries rolled under the impact of the collision and went into an inverted dive, plummeting 14,000 feet before Osborn was able to regain control. He estimated that he was twenty-six minutes from his destination, Okinawa, and doubted that the plane could make it that far, so he cast about for a landing site. The only one within range was Lingshui military airfield on Hainan Island, off the south coast of the Chinese mainland, the base from which the Chinese jets had taken off. The crew followed the standard operating procedure of destroying data and equipment that the US Navy did not care to share with another country. One member of the crew sped up the process by pouring a pot of hot coffee into the disk drives and motherboards.

The second Chinese fighter pilot radioed in to Lingshui airfield for permission to shoot down the US aircraft. The request was denied and the pilot ordered to return to base. The US plane also radioed a mayday distress signal to the airfield, to which the airfield was required by international convention to respond. The Americans repeated their mayday signal fifteen times but were never answered. This would later give China its grounds for arguing that the landing was illegal because it had not been authorised. Osborn decided that he had no choice but to attempt a landing regardless of whether he had clearance or not. His plane was not going to make it to home base. Landing on Hainan Island was going to be enough of a challenge as it was, since the US plane was flying without instruments, had lost control of the flaps on the left wing and was overweight with fuel. The plane touched down on the Lingshui runway doing 170 knots but came to a stop before the runway ran out.

As soon as the plane was stationary, armed soldiers rushed to the runway and surrounded the plane, removing the crew at gunpoint. The Americans were held for eleven days and subjected to illegal interrogation while the two sides engaged in elaborate diplomacy. Only after the

US issued a guarded letter expressing regret for the incident and for the death of Lt.-Com. Wang were the Americans released. China even gave them back their plane after going over it with a fine-tooth comb. Lockheed Martin engineers were permitted to dismantle it, but the pieces had to be shipped on a Russian cargo jet back to Okinawa. It has since been rebuilt in Georgia and put back into service. The US awarded Lt. Osborn the Distinguished Flying Cross. In keeping with a long tradition of deifying military heroes as protectors of the state, China honoured Lt.-Com. Wang Wei with the title of Guardian of Territorial Airspace and Waters.

This incident happened at 22,500 feet in the air, but what prompted it was the ocean below. The rules governing where aircraft may fly over water have not yet been fully codified. They derive largely from the still evolving body of maritime agreements known as the law of the sea. These are the rules regarding what waters belong to whom and what ships may cross them. They also supply the same rules governing airspace. Just as a ship may not enter the territorial waters of another state without explicit clearance, so aircraft are barred from entering the airspace over the territorial water of another state. We can make sense of what happened on Mission PR32 only if we know something about the law of the sea.

Territorial water is recognised as the thin stretch of water that runs along the shore of a coastal state. Traditionally taken to be the distance that a cannonball could be fired from ship to shore, this safety zone was formally set in 1982 at 12 nautical miles. Coastal states can claim full jurisdiction out to that limit. Since the Second World War, however, some states began to push the outer limit of their jurisdiction much further, in order to restrict foreign access to coastal fishing and seabed mineral resources. Eventually a second outer limit was established at 200 nautical miles (230 miles or 370 kilometres) from the shore. Coastal states may assert exclusive economic control over that zone, but that does not give them the right to eject foreign ships, even warships, within the 200-nautical-mile limit – so long as they stay outside the 12 nautical miles. This provision confirms a longstanding right known as 'innocent passage', which allows ships of any flag to transit the coastal waters of another state so long as they do so directly and expeditiously.

What applies to ships on the water applies equally to aircraft above it.

Pilots are expected to ask for verbal clearance when entering territorial airspace, yet the rules of overflight are not set in stone. At the time the planes made contact, the Aries was roughly 110 kilometres (60 nautical miles) south-east of Hainan Island, and therefore well beyond the 12-nautical-mile limit. The American interpretation was that the plane was engaged in innocent passage over China's continental shelf in the South China Sea. China had a right to monitor that passage, but not to impede it or engage in manoeuvres jeopardising the safety of the plane or the lives of its crew. Interception of this sort was harassment. The Chinese view, by contrast, was that the reconnaissance plane was flying over its territorial waters. Entry into Chinese airspace amounted to an infringement of China's sovereignty, and China was fully within its rights to eject the plane.

Curiously, or perhaps wisely, China has never legally argued its right over the entire South China Sea. The claim is unilateral and phrased as a historical right of sovereignty by right of discovery. Chinese mariners first discovered the islands in the South China Sea, this argument goes, and this makes all these islands and the entire ocean surrounding them China's. The legal Latin term for this sort of claim is *terra nullius*, which is to say, 'the land belongs to no one and therefore is mine because I found it first'. This is the historic claim that Europeans made all over the globe from 1492 onwards justifying conquest. Most such lands were far from nullius at the time Europeans arrived, but European law simply declared the existing inhabitants to be savages and therefore without a state capable of exercising sovereignty. There may have been people there, but the land didn't belong to them.

Terra nullius is still available as a legal claim, although there remains almost nothing left on the globe to which this claim could be made. Should China ever attempt to field this claim in international arbitration, it would face some difficulties. Yes, these tiny bits of terra were nullius, but the claim also requires occupation, and these islands were never occupied until China built airstrips in the ocean contiguous to a few of them. Yes, Chinese sources from the fourteenth century indicate an awareness of the tiny islands in the South China Sea, but is this evidence of discovery, or simply of recording what everyone in the region already knew?

At issue is the dignity of states that feel under-dignified. But even more at issue is the promise of undersea oil beds. No one cares about the islands. They care about what's under them. And so Wang Wei fell to his death.

———————————

Seven years later I was in the basement of the New Bodleian Library poring over something the very existence of which I could not have imagined. It lay unrolled before me across two tables that had been pushed together: an old Chinese paper map of the eastern end of Asia. Its size was remarkable, over a metre wide and almost two metres long. The roller at the bottom showed that it was at one time hung on a wall. Hand-drawn in black ink, it depicted the coasts of China and the islands of South-East Asia. The map itself presented an extraordinary panorama. The land was the colour of pale sand, decorated with mountains painted in pale blue and brown, dotted with black ink to suggest trees in the style of Chinese landscape painting and ticked out with tiny blotches of red. Vegetation ran riot across the map – ferns and stands of bamboo, pine trees and elms, irises and aspidistras, even a few orchids. The ocean was flooded in an uneven greenish wash – it would have been blue before the copper pigment oxidised – patterned with cloud-like billows suggesting waves. The cities and ports dotting the map bore labels written in Chinese characters, circled in black ink and edged in yellow. And criss-crossing the ocean from port to port was a tracery of ruler-straight lines showing the courses that ships once sailed – the first map in history to do so on this scale. I was familiar with Asian maps, but I had never seen anything like it. It was beautiful, it was unique: a historical document, a work of art and a mindscape (to borrow the perfect term from map historian Cordell Yee) of how someone once imagined the Asian world looked. Far more than a dry transcription of topographical facts, it animated an entire world. It was perfect.

I was looking at the map that day because David Helliwell had sent me a message earlier that morning suggesting I come over to the library as soon as possible. David has overseen the Chinese collection at the Bodleian Library for as long as I have been a historian of China, so I knew that anything that excited or surprised him would be worth a look.

Something significant had come to light. As soon as I had finished my morning's teaching, I hurried over and found David in his office. He took me down to the basement, where access was restricted, and there lay the map.

The Bodleian is the library of the University of Oxford. It is named after Thomas Bodley, who proposed to the university that it should have a library like all the great universities on the Continent, and that he would build it from the remnants of the manuscript collection stored in a hall over the Divinity School. The library was officially founded in 1602, and two years later Bodley accessioned his first Chinese book. No one in England could make head or tail of a Chinese book, quite literally: the Chinese convention of turning pages from left to right rather than, as Europeans did, from right to left generated confusion over which was the back cover of a Chinese book and which the front, to say nothing of which was the top and which the bottom. But ignorance of the language did not deter Bodley from collecting Chinese books when these became available, or books in any indecipherable language. It didn't matter that his library possessed books without readers. He was confident that some day someone would know how to read them, and that some day something useful in them would come to light. There was no hurry. The books could sit quietly on shelves or in boxes until the time came when they might be of use. Bodley was collecting for the long term.

David had already checked the records and could tell me that the map had entered the Bodleian Library in 1659 as part of a large donation of books and manuscripts from the estate of a lawyer named John Selden, someone of whom I had never heard. Beyond that, he knew nothing. All he could add was that the map had come to his attention because Robert Batchelor, an American historian of the British empire, noting an entry for a map of China in an old catalogue, had called it up. David had gone down to the basement stacks to dig out the long, narrow box in which it had lain undisturbed for close to a century, and had then called me over to take a look. Thanks to Robert's initiative, here it lay before us, three and a half centuries after it had been deposited in the library and, by our guess, roughly four after it had been drawn and painted.

The more I examined the map, the more it troubled me. It just didn't look like any Chinese map I had seen from the Ming dynasty (1368–1644).

It was all wrong. First of all, it covered more space beyond China than any Ming map usually showed. This was a picture not just of what the people of the Ming thought of as home but also of the vast surrounding region that lay beyond their borders, from Japan in the upper right-hand corner to Sumatra in the lower left. The Philippines and Borneo stood where you would expect them, out in the ocean. The familiar outline of the coast of Vietnam was there too, as were the Malay Peninsula and the larger islands of today's Indonesia. Chinese cartography has certain conventions about how to depict these places, usually jammed in and flattened around the essential body of China. Those conventions were starting to change towards the end of the dynasty, admittedly, but no late Ming map looked like this. The first lesson of map history is that maps are copies of other maps. This wasn't a copy of anything I knew.

The map was also wrong in terms of how it balanced the places it depicted. The heart of the map was not China, which is what Ming maps, whether of the region or of the entire world, had trained me to expect. Instead, the centre of the map was occupied by the South China Sea, the now noisily contested zone that then was commercial common ground for every port and state in East Asia and, as the spice trade took Europe by storm around the turn of the seventeenth century, for ports and states as far distant as Goa, Acapulco and Amsterdam. To arrange his map around a vast sea was a most strange thing for the cartographer to do, not just because tradition didn't allow for it but also because there is almost nothing there: a hole in the centre of the map. Rather than letting the landforms dominate the mindscape he had drawn, the mapmaker had pushed the landforms to the periphery and invited us to contemplate the sea.

Finally, and most troublingly of all, the map simply looked too familiar. How we see East Asia cartographically today, how we recognise the shapes of the land and water, is the result of a lot of subsequent history. Our visual idea of this part of the world is necessarily different from what the Ming picture was. This is not because we get it right and they got it wrong. Their vision and ours arise from two different ways of seeing, two different systems of transcribing geographical reality onto paper. These two ways shouldn't generate the same image, and yet here was a Chinese map from four centuries in the past that minimised the differences in time and style that should have rendered the map less familiar than it was at first sight. It was just too perfect. Not only that, but its attention to sea routes was too smoothly suited to the obsessions of our time as China becomes the main supplier of the world's goods and ships its products all over the world. This map charted the commercial world as no map, East or West, had done before. It made complete sense, and yet it made no sense.

That day I learned that it was called the Selden map, it otherwise having no title caption of its own. I had never heard of John Selden, so the connection with the donor was initially of no interest to me. That has changed. The map could have come into anyone's hands, and yet, as I would learn, it came into the hands of one of the authors of the international law of the sea, and indeed the first to argue that states could claim jurisdiction over the ocean – the very claim China now makes over the South China Sea.

The question of state jurisdiction first arose in law in the years immediately following Christopher Columbus's first voyage across the Atlantic Ocean. In 1494 a papal legate brought together the two emerging maritime nations of Europe, Spain and Portugal, at the Spanish town of Tordesillas for the purpose of determining who had claim over which part of the globe. Whatever was *terra nullius* – land that no one owned – would be divided between them according to a line drawn from pole to pole 370 leagues (1,200 nautical miles) west of the Cape Verde Islands off the north-west coast of Africa. Everything to the east of that line 'shall belong to, and remain in the possession of, and pertain forever to' Portugal. So Portugal got Asia, along with a chunk of Brazil, which its treaty negotiators may or may not have known jutted out across the line. Everything to the west of the line – the Americas and the Pacific – 'shall belong to, and remain in the possession of, and pertain forever' to Spain. The treaty also granted each the right of innocent passage through the other's maritime zone.

Tordesillas was conceived as a boundary treaty to resolve a conflict between two states sharing adjacent territory and to provide a framework for co-existence and prevent future conflict. Because it was over water and not land, however, its ramifications could not be limited to the two of them. Its provisions affected every other nation that sent ships to sea, which eventually came to mean pretty much all of Europe. And so it was via a series of objections to Tordesillas that international law slowly took form.

The laurels for founding the law of the sea, and international law more generally, usually go to the brilliant Dutch scholar of the early seventeenth century, Huig de Groot, better known by his Latin name of Grotius. But to my mind that honour belongs equally to John Selden, who, as we shall see in the next chapter, took up the legal challenges that de Groot posed and thereby laid the foundations of a workable law of the sea. De Groot thought the sea could not be subject to claims of sovereignty, whereas John Selden did. Selden did not argue that the South China Sea, in particular, was under any state's dominion – indeed, at the time he would have argued that it wasn't – but he did hold that the sea could be possessed by a state just as land could. This, curiously, is the argument that China makes today.

What John Selden thought about the South China Sea we will never know, but he did think about the law of the sea, which is inarguably a strong visual feature of the map he acquired. His interest in the law of the sea was a specialised variation of everyone's interest, for the sea in his day had the attention of the entire world. As an indication of how wide that interest was, we need simply note the estimate of one European historian that by 1660, a year after the Selden map entered the Bodleian Library, ten thousand European ships were at sea in search of commodities and markets. Not all of them were on their way to Asia, certainly, but many were. There is no way of counting the number of Asian vessels engaged in the same pursuit, but the count would surely go higher than ten thousand. For this was a world in which Europeans were able to sail around the globe and trade between regional economies because sailors who were not European had already created the regional trade networks on which global trade would depend. In the seventeenth century European mariners were the supporting cast in an essentially Asian drama. As yet they enjoyed few technological advantages over their Asian counterparts. As Louis Lecomte, a Jesuit missionary who journeyed to China, testified, Chinese were able to sail the open seas 'as securely as the Portuguese'. He should have known, for he had taken passage on both Chinese and Portuguese ships. From both ends, the world was being rewoven nautically into a unity.

John Selden sensed better than most that he lived in an age undergoing a sea change: new philosophies and new constitutions, new corridors of trade and new forms of wealth, new ideas about the right relationship between the individual and everything taking place around him. Some who had once lived under divinely ordained monarchs now decided that they were ruled by mere men. Where great wealth once had consisted in owning vast tracts of land, now it involved owning vast cargoes of ships. In tandem with these changes, unprecedented arrangements for living in the world were rising at a breathtaking pace from the outdated medieval foundations on which Europe had so long rested. An entirely new structure of law was required to hold up the new order. Selden was the lawyer to do just that.

But John Selden did more than juggle laws and precedents. He was able to anticipate the future because he quarried more thoroughly than any other legal scholar of his era the records of the past – records that in his case ended up including a Chinese map. What he derived from these records were not just precedents for the law of the sea but the foundations of something far more revolutionary: the idea that the purpose of law was to ensure not the power of the rulers but the liberty of the people. His motto was *peri pantos ten eleutherian*: Above All, Liberty. It may seem like a trite phrase – the posturing declaration of a vain young man – but in Selden's case it was a vow to act against the tyrannies of the age, which he himself would experience at first hand when he found himself imprisoned by two kings in succession. It was a vow that led him first into law, then into politics, and finally into the study of Asian languages, including Hebrew, so that he could decipher texts that might enable him to reconstruct liberty as the fundamental human condition.

Selden did not write 'Above All, Liberty' on the map. I would have liked to see the inscription there in his hand, but maps don't have covers and Selden was not about to deface the original by scribbling on it. He left no comment on it, or about it, other than drawing attention to it in his will. And yet, as we shall see, he is intimately part of its history.

———————

Why should anyone, other than those captivated by the history of the seventeenth century, find an old Chinese map in the Bodleian Library of any interest whatsoever? It will take the length of this book to show exactly what we can learn from this single document, and why it matters.

Let me start by asking you to stick a virtual pin in the exact centre of the Selden map. Draw a circle in your mind around that point, giving that circle a radius of one inch. This circle, you will find, sits at the north end of the South China Sea. At the top of the circle, the southern edge of China, is what looks like a long island with a spine of mountains. Its location suggests that this should be Hainan Island, where the Aries made its emergency landing. But the label says something else: Lianzhou. This was a Ming-dynasty prefecture on the mainland north of Hainan, not an island at all. The only actual large island in this area is Hainan, which in the Ming was known as Qiongzhou prefecture. Not Lianzhou. If we look for Qiongzhou on the map, we can find it, but on the mainland, not on this island. The island marked Lianzhou has two other labels, both associated with Hainan. One is Duzhu, Lone Pig Mountain (known today as Wuzhu, Black Pig Mountain). The highest mountain on Hainan, Lone Pig was used by mariners as a fix to tell them they were sailing past the island. The other label says Qizhou, Seven Islands, which is a tiny archipelago off the north-east corner of Hainan where vessels on the coastal route stopped for water and wood in the Ming. So is this island at the top of the circle Hainan Island after all? No, but we'll get to that later.

Now look inside your virtual circle. You will find there a sloping parallelogram containing two lines of Chinese characters. They read, 'shoals for ten thousand li in the shape of a ship's sail'. Beneath it are three characters, 'islet red in colour', and then below that appears another curtain of tiny islets labelled 'reefs for ten thousand li'. A li is half a kilometre and a third of a mile. 'Ten thousand li' is simply a common expression for 'a lot' (the Great Wall, for instance, is 'the long wall that goes on for ten thousand li'). These labels mark the Paracel Islands. The first ten thousand li correspond to what are now called the Amphitrite Group, the second to the Crescent Group. Together they are less than 250 kilometres end to end – well short of the distance that the Chinese figure of speech imputes, although Ming mariners who found themselves swept into this trap of tiny islets and submerged reefs must have felt as though they were caught in a nightmare ten thousand li in length. The Paracels were nothing anyone wanted to claim. They were nothing but danger; the smart navigator steered clear of them.

I have asked you to draw this circle as a kind of portal linking the past and the present. The present we know. It is the world we currently inhabit, in which states and corporations jockey with their citizens, each other and nature itself for ever more wealth and power, in which airmen fall to pointless deaths in the South China Sea, in which the dignity of the nation trumps the dignity of the individual. The past may feel more abstract, more elusive, but this is why I have chosen to put the Selden map at the centre of this book. It means that we will always have the seventeenth century right in front of us as we go forward. That the centre of the Selden map should also be where Lt.-Col. Wang Wei fell to his death is pure coincidence. And yet, if we are willing to treat history not as something dead and gone but as the dimension in which we live, we may find that they have everything to do with each other: that what happens in and above the seas around China has everything to do with the formation of modern nation-states, the corporatisation of the global economy and the emergence of international law – the onset of all of which flags the precise moment when the Selden map came into being.

Today we transit the skies around China more than we sail the seas; but still we transit, and every time we do, we repeat exactly what so many others have done. Our ancestors traded and travelled, migrated and thieved, aided and intimidated, playing along with whoever held all the cards so that a very few could become pointlessly rich and the rest of us might just survive until morning. The world has changed much in the intervening four centuries, but we have changed less. If we have a tiny advantage over our forebears, it is that we can look back, imagine what might have been and understand what wasn't, and why.

Devoting an entire book to a single seventeenth-century map gives us the scope to understand not just a map but the world in which it was drawn. The map is anonymous, which prevents me from telling its history in a straightforward way. I want to get to its origin, but the only place from which I can begin is where it ended up. From there I will move towards China and backwards to the early years of the seventeenth century in an effort to unlock the secrets of when, where, why and how this extraordinary map was made. Our first foray takes us to England under the Stuart monarchy (1603–1714), when the lawyer John Selden collected the map and the librarian Thomas Hyde annotated it. Our second foray takes us to the seas around China during the Ming dynasty (1368–1644), when Chinese and European mariners were building networks of trade that wove the region into a single system of sea routes. Our third foray will take us into the more particular history to which the Selden map must belong, the history of charts and maps leading up to the time of its creation. This, then, is our backward course, from those who read the map to those who acquired it, and to him who made it. With each foray, we will move closer to unlocking some of the secrets of the Selden map; some, but not all. There will secrets the map keeps to itself.

———————

Standing there at the border post, I did begin to wonder how badly my bit of map smuggling could turn out. There was clearly no way I was going to leave China with the map. But then, without it there was never really a question of not letting me leave. It was a stroke of good fortune, not that I was allowed to leave, but that I kept something much more durable than a sheet of printed paper: a memory. The confiscation provided me with a mental bookmark of a day that might otherwise have slipped forgotten into the ocean of days that make up a life. A map for a memory: not a bad trade, really.

I have travelled much and moved often since that day thirty-odd years ago. Had I been able to keep the map, I suspect that I would have ended up giving or throwing it away long ago. One can keep track of only so many things. Even if it were still in my possession, it would now be buried somewhere in the flotsam of files and boxes that make up a scholarly career, no longer of any account. But then, who knows what might have followed? One day long after I was gone, someone might have opened that particular box, pulled out the map and wondered, what is this?

2
Closing the Sea

The only thing we know for sure about the Selden map is that it was deposited in the Bodleian Library in September 1659. Actually, we don't even know that. We have to infer its delivery from the fact that this is when the bulk of John Selden's library was delivered to Oxford. The delivery is recorded by Anthony Wood, a fanatically bookish young man whose 'natural genie', he once observed, obliged him to spend his life reading old manuscripts in the Bodleian Library. Wood doesn't single out the map for particular treatment. Its long, narrow box was simply one among the hundreds that arrived at the Bodleian and that he volunteered to help the Keeper, Thomas Barlow, unpack and sort after they had been carted up from the Thames. Selden had lined up all his books and manuscripts in a reasonably orderly fashion in his house at Whitefriars, London, on what were then the novel installation

we know as built-in bookshelves. The packers had tried to preserve the order they were in, as that was how the collection had been catalogued after Selden died, but the odds are good that it was all in a bit of a jumble when the boxes were unpacked in Oxford. Each volume had to be identified and then carried up to the galleries that ran around the reading-room at the west end of the library – subsequently known as the Selden End – where they were arranged by subject. Chains were then attached to the more valuable items to prevent more from disappearing than had already vanished out of Whitefriars after his death.

This was the biggest single donation of books and manuscripts that the Bodleian would ever receive. By the standards of the day it was an expensive transaction: the chains cost over £25, and the shipping charges ran to £34. The burden of dealing with the enormous task, it was said, drove Barlow to resign his post. In the course of unpacking the manuscripts, Wood discovered the great man's absent-minded habit of not losing his place when his reading was interrupted, for he found pairs of spectacles stuck in here and there among the pages. He dutifully handed in these forgotten bookmarks to Barlow, who reciprocated his thanks by giving him one pair to keep. Wood regarded Selden as the greatest scholar of the age and in his honour preserved this relic of his brush with greatness for the rest of his life.

Selden made his will on 11 June 1654. To it he attached a codicil dealing with the disposition of his scholarly assets (Fig. 2). In this document he singles out his 'Mapp of China'. It is the only reference he ever made to the map. It stands out additionally as being the only item in his collection that he names. The rest of his books and manuscripts he refers to simply by category. He draws notice to his books 'that are in hebrew, Syriack, Arabick, persian, turkish or any other Tongue usually understood by the name of orientall' – these are his manuscripts in Asian languages – but doesn't name any of them. He also mentions his manuscripts in Greek, and 'all such talmudicall and Rabinicall books (if any such I shall have among mine) as are not already in the Library' – that is, his extensive Hebrew manuscripts. The codicil declares that these materials should go to Oxford, although without actually saying so. His 'Mapp of China' he deals with some lines further on. It is to go, he states, to 'the said Chancelor Masters and Schollars' (the 'Chancelor' being the head of the university, the 'Masters' being the professors and the 'Schollars' being the undergraduates). The word 'said' implies that he has already named the university constituted by this body of academics. In fact he hasn't. Something is missing.

After you get used to the crabbed handwriting of the clerk in the Prerogative Court who copied the will – the only copy that survives – the mistake is easy enough to spot. It comes in the fourth line of the paragraph, which reads from the left-hand margin to the right:

> ... understood by the name of orientall or in greek to as also with
> them ...

After much palaeographic head-scratching, I decided that something is missing after the word 'to'. The line should read:

> ... understood by the name of orientall or in greek to *the Chancelor,
> Masters and Schollars of Oxford University*, as also with them ...

In other words, the recipients of the donation fell out of the text. This omission was neither Selden's intention nor his error. The clerk appears simply to have skipped a line when copying the original document, probably when the will was filed and attested in court in February 1655.

The missing line did not puzzle his executors, presumably because they possessed their own, correct copy. Later that same year they released the first batch of manuscripts to Oxford. These were precisely those in the categories named in the codicil: the manuscripts in Greek, Hebrew and the various Asian languages. How the rest of his collection should be disposed of appears from the will to be a decision that John Selden left to his executors. They could 'part the books among themselves, or otherwise dispose of them, or the choicest of them, for some public use', although they should not put the books up for 'any common sale'. He further suggests they might give them to 'some convenient library publique or of some college in one of the universities'. Not until September 1659 did the bulk of his library arrive in Oxford.

The four-year delay excited malicious gossip. It was widely assumed that, as Selden had died without heirs, or at least any whom he chose to acknowledge, his library was destined for Oxford. And yet for four years the books languished in Whitefriars. Some speculated that Selden was thumbing his nose at Oxford, allegedly because Thomas Barlow had turned down his request to borrow some manuscripts from the Bodleian in his last year of life. It is true that Barlow was unhappy about letting manuscripts leave the library, as many were never returned, and he wrote a report to the university saying so. But the university, which to some extent owed its survival to Selden's deft interventions during the worst of the Cromwell years, overrode his objections: Selden could borrow what he wanted as long as he took no more than three items at a time, left a bond of £100 (a huge sum) and returned them within a year. Barlow went up to London nine days after Selden's death to see about the legacy. When the books did not immediately arrive in Oxford, however, tongues started to wag. In the end, the delay was due to complications arising from arranging separate donations of his law and medical books to other institutions. Once those had been sorted out, the rest of the library was free to go to Oxford, and so it did.

There was never really any question of where the bulk of Selden's library should go, and none at all as to the disposition of the Selden map. What eludes us is whether the map went in the first batch of manuscripts, in 1655, or the second, in 1659. It doesn't particularly matter to the story. I mention it only to highlight the fact that this object is not going to yield up easy answers to our questions.

In December 1618 John Selden was ordered to appear before James I. Selden turned thirty-four on the 16th of that month – a Thursday, as it happens. The royal audience took place within the week either side of his birthday. Selden was buried two days short of his seventieth birthday, so in December 1618 he stood one year short of the mid-point of his life. The portrait of him early in his career shows a man who spends much of his time in thought, somewhat apart from the world in which he finds himself (Fig. 3). Ahead of him glimmered a tumultuous career as a constitutional lawyer, a parliamentarian and a legal scholar whose writings would lay the ground rules for modern English law. Ahead of him also lay his private marriage to the brilliant Elizabeth Talbot, the countess of Kent and the widow of his erstwhile patron. No church record of a marriage exists, but that would be characteristic of Selden, who regarded marriage as a civil contract and not a sacrament, and therefore no business of God or of the Church. Or of anyone else, for that matter. 'Of all actions of a man's life, his marriage does least concern other people,' Selden once wryly observed, 'yet of all actions of our life it is most meddled with by other people.' Elizabeth's house at Whitefriars, conveniently located two blocks east of his law office in the Inner Temple, became his after they moved in together. None of this could have been predicted from Selden's modest origins and scholarly bent – and none of it would have happened had James I not called him on the carpet.

London was the place to be if what Selden wanted was to taste the remarkable cultural ferment of the Tudor age. Elizabethan London brimmed with brilliant men from modest backgrounds – theologians, lawyers, poets, playwrights – and John Selden soon found his place among them. His friends included the poet John Donne, the satirist Ben Jonson and the playwright Francis Beaumont. There is no record that he met Shakespeare, but given the circles in which Selden moved – *Comedy of Errors* and *Twelfth Night* were first performed at the adjacent Inn of Court, for example – it would have been difficult to avoid the meeting. Selden's accomplishments later in life would be many, but he was never to make his mark as a poet. Poetry was something every young man of social ambition in Elizabethan London had to write, and so a twenty-something Selden dutifully turned out verses, but verses so laden with obscure references that reading any of them is more like solving a crossword puzzle than reading a poem. The best a fellow poet could say about him with regard to his literary accomplishments was to call him 'solid Selden': praise, if a little faint. Late in life he would declare that verse was 'a fine thing for children to learn to make'. A poet of ordinary talent might write poems 'to please himself, but to make them public, is foolish'. He was probably thinking back ruefully to his younger self.

Selden's closest friend in the literary world was one of the two or three greatest poets of the age, Ben Jonson (Fig. 4). Selden recalled that when James I demanded that Selden present himself for questioning, it was Jonson who accompanied him to the meeting. Here was another who had risen from obscurity on the power of his talent and intellect. The son of a bankrupt preacher who died before he was born, Jonson was raised as the adopted son of a bricklayer: not very different from Selden, whose father owned barely enough land to support his family. Like Selden, Jonson would spend time in the king's prison, and not for what he thought but for what he dared to write. Even though Jonson was twelve years Selden's senior, they spotted each other as kindred spirits in Selden's first year in London. Jonson was not an easy man to please; for Selden to withstand his scrutiny was quite an achievement. When Jonson came out of prison in 1605, Selden was among the guests at the banquet celebrating his release. Despite Jonson's reputation, entirely deserved, for being 'a great lover and praiser of himself; a contemner and scorner of others; given rather to lose a friend than a jest', he was also someone who 'thinketh nothing well but what either he himself or some of his friends and countrymen hath said or done'. Selden was one of those friends. Their personalities were worlds apart, Jonson 'passionately kynde and angry', Selden scholarly and cool. Jonson had come out of the satirical 1590s; Selden was a product of the hedonistic and more polarised 1600s. Yet mutual affection bound them together. Selden loved Jonson's wicked humour, and Jonson knew that Selden was the smartest man in the room. He would always be numbered among Jonson's closest friends, one of 'the Tribe of Ben'.

Jonson was the ideal companion to go calling on a censorious king. It was Jonson who had been commissioned to write the masque welcoming King James and Queen Anne to their new home in 1607, and who had been doing his best to keep the royals amused ever since. The first show was pure Vegas. The text fairly drips with what seems to be craven flattery for 'the fairest queen' and 'the greatest king', whose reign is 'a splendid sun' that 'shall never set'. The central character in the masque is the genius or spirit of the house, who is rather depressed at the thought that his domain is about to undergo a change of ownership. Genius finishes his gloomy opening monologue with the lines: 'And I, uncertain what I must endure, / Since all the ends of destiny are obscure.' Selden may have been in much the same gloomy mood as he and Jonson made their way to see the king. Jonson was familiar with the court, but it was a world Selden had never entered. He belonged among the poets and the lawyers, not the lords of the realm. He was nobody; his future hung in the balance.

Selden grew up in rural Sussex a mile from the English Channel. His father, known as John Selden the Minstrel, was a small farmer who supplemented his income by playing at church services and banquets. His better-born mother, Margaret Baker, claimed kinship with the Bakers of Sissinghurst, a gentry family over the border in Kent. There is no evidence of any truth to this doubtful claim, but it was strong enough for her son to apply for, and be granted, the right to use their triple-swan coat of arms later in life. Young John displayed such brilliance at school that his teachers passed him up from one to the next all the way to Oxford. After four years at the university, Selden abandoned Oxford for London, and the academic life for training in law. In 1604 he was admitted to the Inner Temple, one of the four Inns of Court, where young men prepared to qualify as barristers. Three years before he was called to the bar in 1612, the Inner Temple passed a new rule that 'none should be admitted of this House but those of good Parentage and Behaviour'. One wonders whether the rule would have touched the son of John Selden the Minstrel had he arrived half a dozen years later than he did. And now in 1618 the minstrel's son was commanded to appear before the king.

What brought Selden to the king's attention was his most recent book, *The Historie of Tithes*. It had captured the attention of educated readers, going through several printings in its first year: impressive for a young, unknown scholar. It might be hard for us to imagine how five hundred pages on the history of ecclesiastical tithes could open up a gush of controversy, but this book certainly did so. Its argument was that the right of the Church to collect levies from parishioners was not divinely ordained. These levies or tithes were paid on the basis of a contractual relationship between the Church and the people. God did not command that this be done. Churchmen were appalled: Selden, they felt, had pulled aside the curtain and shown them manipulating levers as though God had nothing to do with what they claimed was theirs. Some of them wanted Selden's head on a figurative platter.

The trouble did not stop there. The buried bomb in the book was the argument that there was no divine right of anything, starting with the bishops but leading ultimately to the king. James I was not the king to welcome this revision. Fancying himself a Renaissance man, James wrote learned essays on political and moral issues that he thought were rather good and that his subjects should read. His particular favourite was *The True Law of Free Monarchies*, a treatise he published in Edinburgh while still James VI of Scotland, then twice in London after becoming James I of England, once upon his accession in 1603 and then again in 1616. Being of a stubborn and unimaginative disposition, James gets into a bit of a tangle trying to justify obedience to tyrannical kings, but the general messages are clear. First of all, kings rule as God's representatives: God appoints them to administer the world on His behalf. Second, in his words, kings are 'the authors and makers of the laws, and not the laws of the kings'. Selden was too thorough a constitutional lawyer not to have read the treatise; indeed he probably read it in both its Scottish and English editions. He must have known that James upheld the divine right of kings. He must also have realised that, if *The Historie of Tithes* came to the king's attention, this would not be good. On the other hand, he probably never anticipated that the king himself would become one of the book's readers.

Selden's defence before the king was the classic historian's plea. He was simply reporting what the historical sources told him, in order to correct erroneous views that had no basis in fact or law. 'Every ingenuous Christian would be glad to know' what he had found, or so he assumed. If erroneous views were left uncorrected, they would eventually undermine the true legal status of ecclesiastical tithes and leave the Church without revenue. He saw it as his duty to set matters straight. His purpose was not to impoverish the Church but to set theological argument aside and place tithes on a firm, because entirely legal, foundation. 'I doubted not at all', he explained to James, 'but that it would have been acceptable to the clergy, to whose disputations and determinations I resolved wholly to leave the point of divine right of tythes.' His intention, he insisted, was to 'keep myself wholly to the historical part'.

It wasn't that simple, and Selden knew it. The fact that he published his study in English instead of Latin is a clear indication that he expected the book to be read beyond narrow academic circles – and that he expected controversy. His challenge to his contemporaries was to understand that the law was an entirely human set of rules. As he put it later,

'Every law is a contract between the king and the people.' The king might legislate on behalf of the Church and thereby require tithes to be paid, but the Church could not legislate on its own behalf, nor could it expect God to. 'There is no such thing as spiritual jurisdiction', he declared. 'All is civil; the church's is the same with the lord mayor's.' Selden did not dispute the legal right of the Church to collect tithes, only its claim that this was the Church's by divine right.

Unfortunately for those of us who would like to eavesdrop on their conversation four centuries later, there is no record of what exactly passed between the two men during the two private audiences in which they met to discuss tithes. According to Selden, he stuck to his views as nicely as possible. He was appropriately sorry that the bishops were distressed by his evidence that God did not give them the right to tax the people, but he implied that it was the bishops' problem, not his. He was just try-ing to correct mistaken assumptions. Fortunately for Selden, James got drawn into the discussion and lost interest in whatever was upsetting his bishops. He seems not to have picked up the potential challenge to his own divine right to levy taxes. He should have, for an attack on the divine right of bishops had only to move its aim by a few degrees to hit the king.

James let Selden off lightly. He banned the book and forbade Selden from engaging in any further discussion of tithes. But it could have gone a lot worse for the young lawyer. Being a friend of James's favourite enter-tainer may have helped. In any event, he did not have to go to jail, at least not this time.

The Historie of Tithes would have gone unremarked had it not been part of a larger sea change in political thinking that was under way. On the one side was a monarchy jealous of its powers, on the other a citizenry increasingly bold about demanding its rights. The thread that would connect Selden to the map that he did not yet own began at his meeting with the king, and quite by accident. Attending James at that meeting was George Villiers, then marquess (later, duke) of Buckingham. Buck-ingham was the king's personal favourite, and he was not shy about using his connection to build for himself a powerful position at court, which included being Lord of the Admiralty. Selden met him for the first time that day. Buckingham treated him well, if we can trust the overly obliging comment Selden made a year later in a letter to the English ambassador to France, in which he mentions the marquess's 'most pleasant behaviour and great humanity towards me (one who was completely unknown to him and completely unused to affairs of court, not to mention undeserving)'. But then, what else does a lamb say about a wolf?

Buckingham's interest in Selden was entirely separate from his master's. Selden was not one to hide his talents under a bushel, and he must have boasted to a few people about something he was writing some time earlier. Buckingham had got wind that Selden had written a rebuttal of a recent Dutch treatise that stood in the way of Britain's ambition to become a maritime power. Buckingham wanted it. The offending treatise had been published nine years earlier, in 1609. The title-page listed no author, but the open secret was that it was the work of Huig de Groot. The brilliant Dutch prodigy, better known by his Latin name of Grotius, was twenty-six years old at the time the book was published, already a veteran of letters and politics. He published his first book at the age of sixteen, on the liberal arts, was appointed an advocate in The Hague the same year and at eighteen became the official historiographer of the States of Holland. He was the rising intellectual star of his generation and would remain prominent throughout his life, most of which he ended up spending in political exile (Fig. 5).

The book that bothered Buckingham was entitled *Mare Liberum* ('*The Free Sea*', usually referred to as '*The Freedom of the Seas*'). In it de Groot argued just that: that no state could exert exclusive jurisdiction over the ocean, and that the ships of every nation were at liberty to sail wherever they chose in pursuit of trade. It might better have been titled *Free Trade*. The particular legal issue that de Groot was tasked to address was the Portuguese claim that, following the papal division of the world between themselves and the Spanish in 1494, the Dutch East India Company (or VOC) had no right to be sending ships into East Indian waters. The book arose in very particular circumstances, and in the service of very particular interests. It was a gauntlet thrown down against the Portuguese, but with enough general force to be a slap in the face of any nation that might seek to block Dutch entry into global trade, and enough legal logic to mark the beginning of what we know today as international law.

The incident that set this dispute going took place on 25 February 1603, at the southern tip of the Malay Peninsula, known today as the Singapore Strait. A VOC captain, Jacob van Heemskerck, had been trawling for spices around the southern edge of the South China Sea for a year without much success. The first port of call for the Dutch and English merchants newly arriving in South-East Asia at this time was Bantam, a small independent kingdom and trading port at the west end of the island of Java. Van Heemskerck had been able to load five ships with spices here the previous spring and send them back to the Netherlands, but his principal goal was to break the stranglehold that Portugal had over spice producers in the Molucca Islands to the east of Java, known as the Spice Islands.

Portugal was pursuing a vigorous campaign to keep the new interlopers from northern Europe out of its trade zone, executing captured Dutchmen as a warning of the lengths to which it was prepared to go to keep its competitors out of this market. Van Heemskerck made no headway in the Spice Islands and went west to Pattani, an international port on the east side of the Malay Peninsula. There he struck up a relationship with Raja Bongsu, the brother of the sultan of Johor, a small regional power that occupied the southern tip of the peninsula and which had declared war on the Portuguese for their high-handed tactics in the region. Van Heemskerck was desperate to come up with some way to make a fortune on this voyage, and Johor was eager to rid itself of the Portuguese, so between them they hatched a scheme to capture the next Portuguese cargo ship that passed through the Singapore Strait. If Van Heemskerck couldn't acquire spices by purchase, he could get them by seizure. The *Santa Catarina* was in transit between Macao and Malacca with a load of gold and merchandise and over eight hundred crew and passengers when the Dutch attacked. After a day's bombardment, carefully designed to incapacitate the ship without sinking it, the Portuguese had no choice but to surrender. Everyone on board was spared and sent on to Malacca unharmed, but the ship and its cargo were taken back to Amsterdam. The profit on this seizure was enormous.

Portugal objected, and the seizure of the *Santa Catarina* became a case before the Admiralty Court of Amsterdam. To no one's surprise, in September 1604 the court found in favour of the plaintiffs – that is,

Van Heemskerck and the VOC. The Company argued that the ship was legitimate booty taken in a just war against Portugal. By the law of nations, both the Netherlands and Johor enjoyed the right to enter into trade relations without being forced to submit to the interests of a third party. By what was termed 'natural law', a sea captain such as Van Heemskerck had the right to punish an offender in the absence of effective justice. Aware that it had won the case on the basis of shaky legal logic, the VOC decided to get a proper legal opinion as soon as the judgement came down. The younger brother of one of the directors had been de Groot's room-mate at university, and this was the connection that led to the VOC's commissioning of de Groot to produce a legal brief on the company's behalf. Drawing on a mass of documentation that the VOC made available to him, de Groot went way beyond his mandate and composed an enormous legal manuscript entitled *On the Law of Prize or Booty*. Chapter 12 of that manuscript dealt with the question of whether the sea was free, and therefore whether the Dutch were justified in applying force against a third party that sought to impair the movement of Dutch ships and interdict Dutch trade with indigenous rulers. As competition with England for trade grew, that one chapter was leaked into publication as *The Free Sea*.

James banned the book as soon as it appeared in England, but he couldn't ban the man. He got a taste of the young Dutchman's style four years later, in 1613, when the Dutch sent him to London as part of an official delegation to discuss trade disputes. De Groot's brilliant mind, excellent Latin and blithe self-confidence in the face of those who disagreed with him – which the English found irritating in one so young – made him the ideal official spokesman at the opening and closing sessions of the negotiations, as far as the Dutch delegation was concerned. James had a particular reason to attend those sessions: he was Scottish. Dutch fishermen had been harvesting herring off the east coast of Scotland for decades, and no English monarch had bothered to intervene. James's aunt, Elizabeth I, had regarded the seas as free and open; it did not occur to her that it was her business to tell the Dutch to quit the North Sea. It would have done no good in any case. The Dutch were better equipped to work on the open seas and relied on the herring fishery as the foundation of the global empire they were just at that moment beginning to build. James viewed matters otherwise. The herring the Dutch caught should have been going to Scottish fishermen, and if they weren't, the Dutch should be paying him a toll for the right to fish there. Now that he was king not just of Scotland but of England as well, he could move the issue to the top of the state's agenda.

In his opening address before the king on 6 April 1613, de Groot talked only about the Dutch position in the East Indies. He explained at length how the VOC had several times been obliged to intervene on behalf of Asian rulers with whom they had preferential trading relationships to save them from 'imminent destruction' at the hands of the Portuguese. He pointed out the enormous expense involved in breaking into the spice trade and suggested that, rather than contend with each other, the English and the Dutch should work out 'a fair partnership'. Piling instance upon example, the long-winded, youthful orator operated the machinery of rhetoric so relentlessly that some of those present judged that he had scaled unmatched heights of tedium. But this was diplomacy, and everybody had to be nice. De Groot said nothing about the herring fishery in his dissertation before the king. He regarded his remit as something else entirely. He was there to advance the interests of his employers, the VOC, in Asia. Still struggling to establish themselves profitably in the East Indies, the Company did not want its English counterpart in the same waters competing with them. De Groot could hardly argue that the Dutch now owned the seas that they had disputed with the Portuguese, but he could point out the enormous costs involved in running a trading operation on the other side of the globe.

De Groot returned to the podium six weeks later to give the farewell address on behalf of the delegation. Mercifully briefer than he had been at his first performance, he acknowledged that no formal agreement had been reached. Rather than admit defeat, though, he proposed two provisional working measures: that neither should act against the other in those places where both were established, and that 'in all other parts of the Indies, both nations extend to each other every possible token of friendship, both freely conducting business according to their wishes'.

James had some interest in the Asian trade, but he was much keener to figure out how to make the Dutch pay royalties on the herring trade. The two sides were so completely at cross purposes that nothing was achieved, either on this or on the return English visit to the Netherlands two years later. But the dispute caught the attention of England's own, slightly younger, version of de Groot. Selden had yet to attain the celebrity that de Groot enjoyed, having barely started down the scholarly path that would bring the two men to the attention of all Europe – and of each other. But he probably harboured the same ambition, which was to use his boundless knowledge of the law to guide the public course of human affairs. Although circumstances placed them on opposite sides of the debate that opened the new field of international law, they became each other's most ardent admirers. As de Groot did not return to England after 1613 and Selden never left the island, they never met. Had they done so, they would have been the two cleverest people in the room.

Selden must have heard about the young de Groot's star turn in front of King James while he was just a junior barrister penning the occasional verse for friends' publications. We will never know exactly what prompted his interest, but Selden got his hands on a copy of *Mare Liberum* and decided to write a rebuttal – also in Latin, of course. The result is *Mare Clausum*, literally *The Closed Sea*, to which the English translation of 1652 gives the clumsy title *Of the Dominion, or, Ownership, of the Sea*. This is the legal treatise that Buckingham hoped would supply him with arguments over who controlled the herring fishery.

Selden didn't immediately give Buckingham the original draft of *The Closed Sea* in 1618 because he wanted to develop the work into something major. *The Historie of Tithes* had run to five hundred pages, and so would *The Closed Sea* by the time its author was done with it. The following summer he submitted a handwritten copy of his enlarged manuscript to James. It was passed from James to Buckingham to the Court of the Admiralty, then back to the king. The timing was not brilliant. Selden argued that British sovereignty extended over the entire North Sea right up to the shores of Denmark – not a little unlike China's current claim to the entirety of the South China Sea. But real-world politics got in the way. It turned out that James owed his wife's brother a considerable sum of money, and that brother-in-law was the king of Denmark. He was in no position to tell his brother-in-law that his subjects had no right to fish in the North Sea. So Selden obediently removed the claim from the manuscript and then resubmitted it, but *The Closed Sea* languished somewhere along the corridors of power, possibly unread, certainly unpublished. Word of *Mare Clausum* got out on the continent, prompting a colleague in Paris to write to its author in 1622, asking whether it had yet been published. Meanwhile royal attention, as it so often does, turned elsewhere.

So did Selden's. One effect of having come to the notice of the king, he discovered, was that he also came to the notice of everyone else. The decade of the 1620s would see much less scholarship in Selden's life and much more politics, as men of influence sought his advice on constitutional issues and gradually drew him into public affairs. He did not hold a seat in the brief Parliament of 1621, but he was engaged as a legal consultant by both the House of Lords and the House of Commons. A popular saying of the time was that the Lords went to Selden to know their privileges, and the Commons to know their rights. The particular issue that drew him into the centre of political controversy and brought him to public attention was the successful attempt by the House of Lords to revive the long discontinued practice of impeaching officials accused of unlawful activity. (The deliberation of the House of Lords to extradite the Chilean dictator Augusto Pinochet to Spain in 1998 rested in part on the powers the Lords re-established for themselves with Selden's advice in 1621.)

King James was not amused. Twelve days after Parliament was adjourned, he ordered three men to be arrested 'for speciall causes & reasons of State knowne unto himself'. One of them was John Selden. He was taken into custody but then released five weeks later without charges being brought. Ben Jonson would have known about the arrest immediately. The poet laureate was not, however, in a position to abandon the benefits he got by pleasing the king, especially when he found himself having to sign away his pension to a creditor the same month Selden went into prison. Selden needed relief, but Jonson needed money. Buckingham came to the rescue, paying him an advance of £100 to write a masque for James's visit to his new country house. Within a month Jonson dashed off *The Gypsies Metamorphosed*, his longest masque, casting Buckingham in the starring role as the Gypsy Captain. It was an audacious gesture. Gypsies in Stuart drama were figures whose freedom placed them above the moral constraints enchaining ordinary people. Turning Buckingham into one hinted at his pariah status as the king's favourite. Some readers of Jonson go even further and suggest that Jonson was hinting at Buckingham's reputation as James's lover.

If we put John Selden in the picture, however, the picture changes. In the first speech in which the Gypsy Captain/Buckingham addresses the king, he praises James for following a foreign policy that eschewed war as a tool for settling the conflict on the Continent between Catholics and Protestants:

> For this, of all the world, you shall
> Be styled JAMES THE JUST.

The capital letters are in the original. James the Just? Selden, the friend whom Jonson praised that very year as 'the Law book of the Judges of England', was in detention. Was anyone listening?

Jonson cuts just as close to the bone in a later scene, in which the Third Gypsy describes a troop of drunken soldiers who declare that they have a right to steal food,

> As being by our Magna Carta taught
> To judge no viands wholesome that are bought.

The Great Charter was the revered touchstone of the rights of subjects towards their rulers. Originally written in 1215, it was the backbone of the ancient constitution, which insisted, among other things, that 'no freeman shall be imprisoned but by the law of the land'. Selden owned half a dozen manuscript copies of Magna Carta and cited the text regularly in his writings. Jonson would have known all this. So who was it who was really joking with the law? Luckily for Jonson, James had a tin ear for political rhetoric when it was disguised as verse. He was too delighted by the masque to notice the barb.

By the time the masque was performed, however, Selden was out of prison. The crisis was over; hearts could be light again.

The arrest had been a warning to Selden, but one he declined to take; indeed, it had the opposite effect. Rather than feel chastened, the short spell in prison galvanised him to oppose anything his king tried to do 'for speciall causes & reasons of State' that impaired the rights of citizens. He would spend the next three decades countering every such attempt to use extra-judicial means to force compliance, whether by the king or, even more radically, by Parliament.

In the next Parliament, of 1624, Selden did not just advise the House of Commons but sat as a member. Before controversy could pit Parliament too squarely against James, the 'lord of the four seas, king of the less and greater isles', as Jonson once called him, had the good grace to die. His son was crowned Charles I. The arrogance and ambition of the new king put him on a collision course with Parliament right from the start, and Selden was inexorably drawn into the fray. Buckingham (now the duke of) continued to be the son's favourite as he had been the father's. Disliked under James, he was loathed under Charles, and Parliament set its sights on bringing him down. Selden was on the committee in 1626 that drew up articles of impeachment against him. Charles came to Buckingham's defence by adjourning Parliament. When it reopened in 1628, Selden was again involved in preparing the text of the impeachment. Before preparations were much advanced, a disgruntled soldier made the impeachment unnecessary by assassinating the duke in a tavern.

The final session of Parliament in 1629 ended in uproar over the refusal of the House of Commons to grant Charles his request to collect 'tonnage and poundage'. This was a tax on commercial vessels importing commodities (called 'tonnage') or exporting them ('poundage'). Constitutionally the king could not levy taxes without the assent of Parliament. But Parliament was in no mood to accommodate until the king agreed to acknowledge the Petition of Right. This was a legal measure by which Parliament denied the king the right to imprison someone without a charge or, to go back to Selden's first arrest, to do so claiming 'speciall causes & reasons of State knowne unto himself'. Selden was one of the Petition's authors.

After shutting down Parliament once again, Charles I did as his father had done. He had the parliamentarians who offended him arrested. This time there were nine of them, and once again Selden was among their number. His scholarly reputation on the Continent was growing, and observers from Huig de Groot to the painter Peter Paul Rubens expressed outrage at his arrest. No charges were laid, as whatever the attorney-general proposed, the judges rejected. The wheels of pseudo-justice ground slowly; it took eight months for the nine men even to get a bail hearing. The chief justice who presided over the hearing assured them that the king would grant them all bail as long as they agreed to sign a bond of good behaviour. Selden regarded this as an illegal condition of their release and refused. 'We demand to be bailed in point of Right', he stated, not as a gesture of royal favour. Back to prison they went. Selden was eventually moved from the Tower to easier conditions in Marshalsea Prison, and passed most of his second year on something closer to house arrest. No charges were ever laid. Not until May 1631 was he granted bail, although he remained on probation for another four years. His portrait as a man in later life probably dates to the time of his release (Fig. 6). The face of the young scholar in the earlier portrait has given way to the determined look of a man who has found the measure of his own brilliance and has learned the costs of commitment to the political high road.

The reason for his release from probation in 1635, and possibly for his earlier release from prison, was *The Closed Sea*. While Selden had been out of circulation, Charles I had been ramping up his campaign to assert his sovereignty over everything he could claim, including, once again, the North Sea. The Dutch had effectively monopolised the herring fishery, and Charles wanted to push them out. This required a stronger navy, and so Charles, like his father, demanded a new tax to build naval vessels, called ship money. He also needed legal justification. As early as 1632 he was asking for 'some public writing' that would affirm his sea rights. Apparently through the mediation of the newly appointed Archbishop of Canterbury, William Laud, who was always ready to fix whatever he could for his king, Selden was approached with a deal. In return for publishing *The Closed Sea*, he would be given his freedom. Argue for enclosure, and his own enclosure would end. He took the offer.

Selden had already been working on his treatise off and on through the latter part of the 1620s. By the time his bail was rescinded in February 1635, it was more or less finished. Word of its imminent release got round quickly. De Groot heard about it in May, the Pope in June. The final version was in the king's hands by August, and in November it was out, and in the most expensive edition of any of Selden's books. It may have been a deal with the devil, and the fact that the sole surviving copy of the order for his release from bail in the State Papers is in his own hand hints at just how deep the deal ran. But Selden had been pressing hard for his release even before the deal was struck. In his own eyes it was long overdue. In any case, while he was devoted to the principles of liberty and justice, Selden accepted the value of mixing principle and practice. As he once said in another context, 'In a troubled state we must do as in foul weather upon the Thames, not think to cut directly through, so the boat may be quickly full of water' – there was only one bridge across the Thames in Selden's day, obliging most people to be ferried across in small skiffs – 'but rise and fall as the waves do, give as much as conveniently we can'.

He didn't altogether convince friends or enemies that he had not prostituted his legal scholarship for the 'convenience' of buying his freedom. But these were difficult times in a troubled state. The certainties of the Elizabethan world, held together by fixed moral values and a good dose of state surveillance, had given way to doubt, disorder and their bastard offspring: a revolutionary zeal in matters of religion and politics equally. Selden preferred slow revision to sudden change. Forcing change 'is dangerous, because we know not where it will stay; it is like a millstone that lies upon the top of a pair of stairs; it is hard to remove it, but if once it be thrust off the first stair, it never stays till it comes to the bottom'. Not good for those who happened to be standing in its way.

The 1635 edition of *The Closed Sea* is a remarkable work. It consists of two parts, each of which explicates what Selden calls in his preface his two propositions: 'the one, that the Sea, by the Law of Nature or Nations is not common to all men, but capable of private Dominion or propertie as well as Land; the other, that the King of Great Britain is Lord of the Sea flowing about it, as an inseparable and perpetual Appendant of the British Empire'. Insisting that he will argue for nothing more but nothing less, he then devotes five hundred pages to compiling a detailed defence of Britain's jurisdiction over its surrounding seas in theory and in practice. The book is not entirely objective history. As Gerald Toomer, the pre-eminent authority on Selden, has put it a little acidly, *The Closed Sea* 'is a lawyer's brief rather than a historical treatise, despite its truly impressive exhibition of knowledge of both original sources and modern literature. One might wish that Selden had treated the history of the claims to control the seas surrounding Britain as he had treated the history of tithes.' Toomer finds some of Selden's arguments 'so patently weak or absurd that it is difficult to believe that he himself gave them credit'. He crowns his judgement by quoting from de Groot: 'jurists who use their proficiency in the law to please those in power usually are deceived or themselves deceive.' And yet de Groot praised Selden as 'that humane and learned man' who 'treated me both humanely and learnedly', while Selden championed de Groot as 'a man of great learning, and extraordinarie knowledg in things both Divine and Humane; whose name is very frequent in the mouths of men every where'. This doesn't sound like the language of mortal enemies. And in fact, they weren't. Both men regarded liberty as the natural condition of humankind, and both maintained that it could be constrained only by agreement and never by unilateral imposition; which means that both opposed state tyranny and understood that law was the means to do it.

The Free Sea may be just as burdened by its polemics as is *The Closed Sea*. Both were lawyer's briefs written for their clients, after all: one for the VOC, the other for Charles I. Their difference had mostly to do with the interests they served rather than with the law each sought to uphold: a difference of degree more than of substance. Their audience preferred to polarise their positions, English and Dutch each claiming that their side won this battle of the books, but in fact neither prevailed completely to the exclusion of the other. This is why the international law of the sea today consists of a mixture of the two that recognises both freedom of movement and reasonable jurisdiction. Together the two men co-created maritime law. The next generation understood this explicitly and read both authors.*

---

*Samuel Pepys, secretary to the Navy Board and therefore concerned with such matters, spent his evenings in the winter of 1661–2 reading the two books side by side. He read Selden in the English translation of 1652 by Marchamont Nedham, in which the original dedication to Charles I was replaced by a paean to Parliament's 'Right of Soveraigntie over the Seas'. On 17 April 1663, with Britain once again under a monarch, Pepys had his copy rebound with a new title page dedicated to the king, 'because I am ashamed to have the other seen dedicated to the Commonwealth'. Four days later he wrote in his diary: 'Up betimes and to my office, where first I ruled with red ink my English "Mare Clausum," which, with the new orthodox title, makes it now very handsome.'

Had the Selden map ended up in de Groot's possession, no one would need to guess why. The Singapore Strait, where the original incident over which the VOC hired de Groot took place, sits in the foreground of the map. Had he possessed the Selden map rather than the two European maps the Company lent him, he could have mustered it to support the two foundations of his argument against the Portuguese attempt to exclude the Dutch from Asian waters. First of all, the region was not *terra nullius*, or unoccupied territory. 'The East Indian nations in question are not the chattels of the Portuguese,' he writes, 'but are free men and *sui juris*' – that is, under their own laws. Portugal had no claim as a sovereign power in the region. Secondly, de Groot could use the map to underscore his point that 'the Arabians and the Chinese are at the present day still carrying on with the people of the East Indies a trade which had been uninterrupted for several centuries'. By entering Asian waters, the Portuguese were simply trading alongside the Asians already there. Both circumstances ruled against the Portuguese position that they could exclude the Dutch.

Could Selden have used the map to prove the opposite view, that Portugal had dominion over these waters? The question is hardly worth asking, since this is not what *The Closed Sea* is about. It argues instead, at a far more general level, that the ocean is capable of being placed under state dominion under certain conditions. That dominion did not rule out innocent passage, but it did deny the contention that the ocean was a free space. Arguably, Selden could have used the map to illustrate this point by showing how maritime trade routes concentrated at the nodes of port cities – to illustrate the point but not argue it, since the map is basically indifferent to the legal concerns that excited Selden and de Groot. As we shall see, that wasn't what the Selden cartographer cared about.

Had Selden written something that tied his map to issues of maritime sovereignty, we would be in a better position to say why he wanted to own it. But he didn't. So we may do better to consider a different possibility. Whatever he might have conceived about sea law from the map, his ownership of it sprang from a different motivation. Telltale evidence of that motivation lies in *The Closed Sea*. Flick through its pages and you will notice, especially in the historical section in Book 1, extensive quotations in Hebrew and Arabic script. Selden doesn't just translate them or transcribe their sounds into Europe letters, but has them typeset in the original script. *The Closed Sea* has the distinction of being the first English book to print Arabic in metal type. Selden designed and had the types cast specifically for this project, and at considerable expense. Historians of printing know them as the 'Selden types'.

The book's citation of Middle Eastern languages points to a facet of John Selden we have barely considered. He was the era's most important legal historian and constitutional theorist, but he was also its greatest scholar of the humanists' newest and most challenging field: Oriental studies. Renaissance humanism rooted itself in a knowledge of Latin and Greek. Oriental studies arose to bring the study of languages east of Greece into scholarly practice. The first Oriental language to tackle was Hebrew. Selden was not alone among English scholars in being able to read it. His first instructor in Hebrew was his mentor and great friend James Ussher, later Archbishop of Armagh. Ussher is now best remembered for dating the creation of the world precisely to the night before Sunday 23 October 4004 BC – for which he claimed to find corroboration in Hebrew sources. The serious scholars could read Hebrew, although many could not. The pressure was such that Queen Elizabeth's adventurer Walter Raleigh felt he had to apologise to readers for citing Hebrew passages in his *History of the World* of 1614 without actually knowing the language. He admits this to be a poor showing on his part, given – as he says in one of the funniest lines in the book – that he had had 'eleven years leisure' in which to write it. James I had clapped him in the Tower shortly after coming to power in 1603, and Raleigh was still languishing there when the book came out (he was dead by the time Selden got his turn in the same prison).*

---

*Raleigh follows this apology in his preface with the second funniest line in the book, explaining why its history does not extend to modern times: 'To this I answere, that whosoever in writing a moderne History, shall follow truth too neare the heels, it may haply strike out his teeth. There is no Mistresse or Guide, that hath led her followers and servants into greater miseries.' Raleigh was released in 1616 but executed in 1618 for failing to redeem himself in James's eyes. Ben Jonson

Selden read and wrote Hebrew with ease, but what distinguished him from most of his contemporaries was that he also read several other Middle Eastern languages. In his will he mentions Syriac and Arabic (he started both under Archbishop Ussher) as well as Persian and Turkish. His linguistic competence arose out of his practice as a historian. The best way to untangle a controversy – especially religious controversies, which invariably got tangled up in bad logic and worse claims – was to go back to sources as near as possible to the historical period under examination. Some people might think they can dismiss this sort of research as 'the too studious affectation of bare and sterile antiquity', which ended up appearing to be 'nothing else but to be exceeding busy about nothing', as he writes in the opening dedication to *The Historie of Tithes*. In fact, much was at stake. Being able to draw on documents from Oriental traditions simultaneous with the Biblical tradition – and being able to read them in the original language – gave the informed scholar new tools to break old puzzles that remained locked as long as they were looked at on their own terms and not from the outside.

The mention of being 'busy' is a sly nod to *Bartholomew Fayre*. Ben Jonson set his play, written four years earlier, at the notorious cloth trade fair held every August in the parish of St Bartholomew the Less. According to one excited pamphleteer, the carnival attracted

> people of all sorts, High and Low, Rich and Poore, from cities, townes, and countrys; of all sects, Papists, Atheists, Anabaptists, and Brownists; of all conditions, good and bad, vertuous and vitious, Knaves and Fooles, Cuckolds and Cuckoldmakers, Bauds and Whores, Pimpes and Panders, Rogues and Rascalls, the little Loud-one and the witty wanton.

It was the perfect setting for Jonson to mock the ignorant and self-righteous by having them laid low by the quicker wits of those who had no pretentions to virtue. In a late scene in the play a recently converted Puritan zealot by the name of Rabbi Zeal-of-the-Land Busy condemns

---

tutored Raleigh's free-spirited son Wat on his European tour in 1612 – disastrously, it is said, as he was drunk much of the time.

the wooden puppet Dionysius for being an idol; Catholics were idolaters, good Protestants were not. The two get into a roaring argument about whether being an idol is a profane occupation. Busy then moves on and chides the boy puppet for being an abomination on the grounds that he can dress as a male or a female, depending on the parts he plays. This refers to a popular Puritan canard that cross-dressing is condemned in Deuteronomy 22:5. Dionysius dismisses it as the 'stale old argument' against male actors cross-dressing to play female characters on stage, 'but it will not hold against the puppets; for we have neither male nor female amongst us. And that thou may'st see.' The script then directs the puppet to lift his garment, flashing his non-existent genitals. The joke was on Busy.

Selden loved the joke, and returned the favour two years later by writing for Jonson a deliriously academic treatise on the passage in Deuteronomy that Puritans liked to cite to attack the theatre. He drags Jonson through a bewildering lattice of texts in Latin, Greek and Hebrew in what is both a perfect example of informed antiquarian scholarship and a complete send-up of Puritan intellectual incompetence. The point he drives into the ground is that the passage means nothing of the kind. What was originally a reference to the Hebrew God's condemnation of rites performed to bisexual deities such as Venus and Baal – rites in which women donned male armour and men wore women's robes – was taken out of historical context to prove that God condemned cross-dressing. It is a historically inaccurate interpretation, Selden demonstrates, although he ducks out of religious controversy at the end of his treatise by wryly declaring, 'I abstain to meddle.' But he has meddled, and the lesson is crystal clear. Theologians who fail to go back to the original sources shouldn't use scripture to harass people they don't like. The antiquarians were doing their work. It was the Puritans who were 'exceeding busy about nothing'.

The point here is that fluency in Hebrew and other ancient Middle Eastern languages was the new methodology for serious history. It had been part of the new methodology that resulted in the King James Bible, yet in the hands of scholars like Selden it would become a tool to discover flaws in its text. As Oriental studies changed the rules of scholarship in all fields of historical and legal study, the ability to read Asian languages became the cutting edge for the creation of new knowledge in the humanities. Selden's last great scholarly project, on the early history of the political constitution of the ancient Jews, might look from a distance like just another case of 'bare and sterile antiquity', but that was not what he was up to. He delved into this lost tradition for the explicit purpose of uncovering the constitutional principles that should underpin the constitution of Parliament. In Stuart England the smart people were reading ancient Oriental languages. The fact that Europeans were moving into the contemporary Orient in ever greater numbers at this very time only made this curriculum more compelling.

So perhaps it wasn't only the law of the sea that prompted Selden to acquire a large Chinese map. He was certain that every manuscript encoding Oriental knowledge had the potential to reveal world-changing knowledge, and should therefore be collected and preserved, even if no one could yet make sense of it. Although the poet John Milton declared Selden 'chief of the learned men reputed in this land', and the popular Welsh writer James Howell in 1650 declared, 'Quod Seldenus nescit, nemo scit' – 'What Selden doesn't know, no one knows' – not even John Selden could read Chinese. But he didn't need to read Chinese to grasp the geography of the map. It would have been no challenge for him to pick out the spot where, for example, Van Heemskerck captured the *Santa Catarina* at the south end of the Malay Peninsula in 1603. He couldn't have read the place-name by which Chinese knew Johor, but he should have been able to construe the lines that go in and out of Johor's harbour and connect it to the main shipping route connecting the Gulf of Siam with Melaka.

Take a good look at this bit of the map and you will discover that just offshore, right where the sea route veers into port, the surface has been abraded more than any other place on the map. The label for Johor is still legible, but the shipping routes have been almost entirely worn down to the bare paper. Is this damage simply random wear and tear? Or might it be a tell-tale sign that this was the location that most interested its owner, who liked to point it out to friends? Is this the one mark that Selden – unwittingly, like his glasses – placed on his map?

As for reading Chinese, Selden wasn't flustered. It didn't matter. For now it was important just to collect manuscripts, to add materials in whatever languages to England's reservoir of knowledge so that future generations might discover what the present generation could not yet learn. Some day someone would be able to read them and unlock the secrets they contained. That was what mattered. It would take twenty-eight years after the map was deposited in Oxford before someone arrived who could read Chinese.

3
Reading Chinese in Oxford

Disfigured men, women and children, their necks misshapen by swelling lumps and open lesions, tramped into the city from all over Oxfordshire in the early hours of 5 September 1687, making their way to Christ Church Cathedral. We know the disease as scrofula, although, as antibiotics have made it rare to the point of being non-existent for us, I had no idea what scrofula was until I looked it up. (It is a form of tuberculosis that attacks the lymph glands in the neck.) They knew their affliction as the King's Evil. They also knew that there was only one cure. They had to be touched by a king. Whatever the faults of the crowned heads of Europe, and they were many, they did not disregard their duty in this matter. Charles I, in whose downfall John Selden played a part, had touched for the King's Evil. So did his French nephew Louis XIV, his elder son Charles II and, last of all, his second son, James II.

Word got out that James II was coming to Oxford that weekend to browbeat the fellows of Magdalen College into accepting his preferred candidate as their new president. The fellows of Magdalen had resisted his choice for five months, forcing James to descend into the political fray and come to Oxford to push his man through – although the fellows would in the end succeed in thwarting the king after he left. It was a poor tactical move on his part. What was supposed to have been a display of royal authority only tarnished his already wobbly reputation, further tipping the scales against him. (What the anti-monarchists called the Glorious Revolution would force him off the throne and out of the country the following year.) The disfigured victims of the King's Evil crowding into Christ Church that wet Monday morning had no interest in monarchical politics. All they wanted was for James to show up for the service and dispense his favour by touching them. Surely that was not too much to ask of their king? And so he did, from eight to almost ten that morning.

A late breakfast awaited him over at the Bodleian Library. Perhaps to mollify him for Magdalen College's refusal to bend to his will, the university laid on a feast fit for, yes, a king, consisting of 111 dishes and costing £160. Even today one diner couldn't run up a breakfast tab that high. The meal was laid out on a large table set up in the Selden End, the western portion of the library, where Selden's magnificent collection of books and manuscripts was arranged on open shelves. Dining at the Selden End was reserved for the most august visitors to the university: James's brother Charles II had been given the same treatment when he visited Oxford in 1663, again after touching for the King's Evil (Fig. 8).

James entered the library from the east side and paused between the pair of great globes, one terrestrial and the other celestial, commanding that end of the library to receive a Latin oration of welcome and to allow his hand to be kissed – the king touched rather than touching. He then turned to the terrestrial globe and pointed out to one of his courtiers 'the passage between America and the back of China, by which certaine ships had passage'. This report comes from Anthony Wood, the bookish young man who found Selden's spectacles in his books. Three decades later he was still haunting the Bodleian Library. The passage the king pointed out was the famous route that Spanish galleons took to carry silver from Acapulco west across the Pacific to the San Bernardino Strait in the Philippine Islands, through which they passed to reach the Spanish base at Manila. The return route along a more northerly latitude was used to transport coveted Chinese manufactures eastwards back to New Spain, completing the circuit that was arguably the driveshaft for the world economy of the seventeenth century. Did James draw attention to the China trade to display his knowledge of the world in general, or was it to make reference to something more particular? He did not elaborate, but simply proceeded to the Selden End and took his seat in the chair of state.

A large table laden with delicacies lay before him. He sat down and sampled the feast while a 'rabble' – this being Wood's contemptuous term for the crowd hovering around him – looked on with envy. After three-quarters of an hour of eating and drinking – he particularly commended the wine – the king got up and left the table. Immediately those in attendance launched themselves at the banquet table to grab what they could. The scholars were quicker than the courtiers. The fastest grabber was a physician from Magdalen College. 'Noted here for a scrambler', he was 'so notorious', Wood wrote in his diary, that the other scholars who wanted a share of the food 'flung things in his face'. There followed what can only be described as a food fight, the effects of which left many a woman's dress covered in dessert stains.

Unable to make his way through the throng, James paused to watch the mêlée for a few minutes. Only then did the crowd step back to open a path for him towards the exit. As he was about to leave the library, he noticed the chaplain who had preached at Christ Church the day before and praised him to the vice-chancellor and the senior academics. The theme of his sermon, the king recalled, was the sin of pride: it was a message they should take to heart. 'There are a sort of people among you that are wolves in sheep's clothing', he warned. 'Beware of them, and let them not deceive you and corrupt you.' Having warned them about the perils

47

of Protestant extremism, James walked out of the library for the last time. Fifteen months later the Catholic monarch fled his realm to seek refuge with his French cousin, never to return to Oxford – or England. The wolves got their way.

Were it not for the food fight, Anthony Wood might not have bothered to record the details of James's visit to the Bodleian or the interesting conversation that took place after he finished his breakfast but before getting up to leave. As he sat there, James turned to the vice-chancellor and asked whether the library had a copy of a recent Jesuit translation of the writings of Confucius. As the Keeper of the Bodleian was on hand, the vice-chancellor suggested that Thomas Hyde would be better equipped to answer him.

'Well, Dr Hyde, was the Chinese here?'

'Yes, if it may please Your Majesty,' Hyde replied, 'and I learned many things of him.'

'He was a little blinking fellow, was he not?' This seemed to be the best way the king could describe the unaccustomed look of epicanthic folds, possibly the first he had ever seen.

'Yes, Your Majesty.' Hyde then ventured to explain: 'All the Chineses, Tartars, and all that part of the world are narrow-eyed.'

'I have his picture to the life' – that is to say, a portrait painted from life – 'hanging in my roome next to the bed chamber.' There being nothing that Hyde could say on the subject of the art the king hung on his walls, he remained silent. James then turned back to the subject of the book he had mentioned to the vice-chancellor.

'It is a book of Confucion translated from the China language by the Jesuits, four in number', he explained to Hyde. 'Is it in the library?'

'Yes, it is', Hyde assured the king. 'It treats of philosophy, but not so as that of European philosophy.'

'Have the Chinees any divinity?' inquired the king.

'Yes', replied Hyde, 'but 'tis idolatry, they being all heathens. And yet they have in their idol-temples statues representing the Trinity, and other pictures, which shew that antient Christianity had been amongst them.' With this observation Hyde alluded to the Jesuit interpretation of early Chinese history, which argued that the ancient Chinese had a primitive knowledge of God and only lacked Christian revelation. It was a controversial position, especially with Protestants, but the comment seemed to agree with James, who nodded. There the king ended the conversation, although when he glanced up at the manuscripts on the shelves high above him in the gallery, Hyde was so bold as to inform His Majesty that they were the donation of the late Archbishop Laud – whom we will meet in a later chapter. Shortly thereafter the king rose, and the food fight began.

The 'little blinking fellow' who came up in the conversation between the king and the librarian was Shen Fuzong, as the name would be rendered today. Hyde wrote of him as Michael Shin Fo-Çung, placing his Christian name on one side of his surname and his Chinese given name on the other. It was more than James could be bothered to remember. No wonder he resorted to calling him the 'little blinking fellow'. A king couldn't be expected to keep all that in his head. Hyde used 'Michael' when he spoke to him, but Shen, conscious of their difference in ages – Hyde was fifty when he met Shen and Shen not yet thirty – addressed his elder as Hyde, never as Thomas. I shall do the same.

Michael Shen's appearance at the court of James II is part of a longer tale, heroic in some eyes and reckless in others, that tells of the missionaries of the Society of Jesus, known as the Jesuits, going to China to convert its people and its emperor to Christianity. The Jesuit project to convince Chinese to give up their indigenous beliefs for Christian ones had been under way for a century between the entry of the first Jesuit into China and Michael's arrival in Oxford in 1687. Michael's connection to the mission was a Flemish Jesuit by the name of Philippe Couplet. Couplet arrived in China in 1659 after a harrowing voyage by land and sea that left the leader of his team, the senior Jesuit cartographer Michael Boym, dead on the border between Tonkin (Vietnam) and China. Having arrived safely in China, Couplet was despatched to work in the Yangzi Valley, where he travelled extensively for the next two decades – except for the second half of the 1660s, when all Jesuit missionaries in China were banished to the southern city of Canton. It may not have been until after the ban was lifted in 1671 that Couplet met a medical doctor by the name of Shen in Nanjing. Couplet inspired the doctor's son to study the

Christian faith, gave him his Christian name and diverted the boy's life course by taking him to Europe.

The round-the-world journey that would carry Michael Shen from China to Oxford started on 4 December 1681 in Macao, where they boarded a Portuguese ship bound for the Dutch trading centre in Jakarta, which the Dutch called Batavia. Couplet was fifty-eight at the time and Michael about twenty-three. Michael was supposed to be one of five Chinese acolytes to be taken to Europe, but in the end only two Chinese set out, and one lost heart for the adventure during their long layover in Batavia. It took seven and a half weeks to get to Batavia. The season for sailing west across the Indian Ocean on favourable winds was not quite over, but the last ship to Europe that year had already departed. There was nothing for it but to sit tight and wait for next year's sailing season. When they met another Jesuit father heading in the opposite direction, the second Chinese student travelling with Couplet and Michael asked to return home with him. Couplet had no choice but to let him go. Michael elected to stay with his teacher. At long last, over a year later, they secured passage on a Dutch ship going back to Europe. A year and a week later they were in Antwerp.

Over the next four years the two were situationally inseparable. Michael needed Couplet to guide, protect and support him as they moved from Jesuit house to Jesuit house. Couplet had just as much need of 'the Chinese convert', as he was called. Michael Shen was a huge curiosity to Europeans, none of whom had seen a Chinese before. He was an exotic specimen of a distant land that only the Jesuits had access to, and his notoriety in turn gained them ready access to courts and polite society all over Europe. As his Latin, Italian and Portuguese improved, so did that access. All doors opened to Michael; and when they did, they stayed open for his handler too. Poised, polite and sufficiently educable to move with ease between the literary and cultural worlds of China and Europe, Michael was living proof that the Jesuit mission was on the right track. The Jesuits had drawn this ideal Chinese to the Christian faith. With sufficient financial backing, might they not do the same with every Chinese? The sky was the limit. And so Michael, elegantly dressed in green silk and deep blue brocade worked with dragons, was put on show for the crowned heads of Europe. Louis XIV was so charmed when Michael visited him in 1684 that he invited him back for dinner the next day just to watch him handle chopsticks.

Then there was the Pope. Would Couplet have got Alexander VII's attention had he not had Michael in tow? Couplet had particular reason to see the Pope, since his main goal in returning to Europe was to ask him to allow the liturgy for Mass in China to be sung in Chinese rather than in Latin. It was a tough sell. This was a pope whose theological achievements included reconfirming, against Galileo, that the earth did not revolve around the sun, and declaring that Mary not only conceived Jesus without human insemination but was herself conceived by virgin birth – a view that James I derided when he visited the Bodleian Library in 1605, wishing that such objectionable views 'could be altogether suppressed rather than be tolerated to the corruption of minds and manners'. Couplet's fears proved well founded. Alexander VII was not about to let the word of God undergo translation into any language, and certainly not into Chinese.

Couplet also had concrete tasks for which he needed Michael. The Jesuits who were confined to Canton through the latter half of the 1660s had decided to make use of their enforced inactivity by organising propaganda for their mission back in Europe. One project was a massive 'historical description' of China translated from Chinese into Latin. Couplet wrote to Joan Blaeu, son of Willem Blaeu – whose family rivalled and then supplanted Jodocus Hondius as the major cartographic publisher in Amsterdam and produced the globes that James II fingered at the Bodleian Library – to convince him to take on this project. Blaeu hesitated. He was unwilling to shoulder such an expensive publication, which would in any case have competed with his earlier *Atlas Sinensis*, the work of the previous generation of Jesuits. Couplet brought a copy of Luo Hongxian's 1555 atlas of China to assist with the maps that would illustrate such an encyclopaedia, but the project never went forward.

The other project they took up during their Canton confinement was the translation of the seminal four texts of the Confucian tradition, known as the Four Books: two short treatises entitled *The Great Learning* and *The Doctrine of the Mean*, the *Analects* or sayings of Confucius, and the writings of his follower Mencius. Only the first three were completed, but that was enough for Couplet to take back to Europe and see through to publication as *Confucius Sinarum Philosophus* ('Confucius, Philosopher of the Chinese'). Although four authors are named on the title-page, as James II noted, the book was mainly the work of Philippe Couplet.

Confucius was the most ambitious scholarly work on China the Jesuits would produce that century. Not merely another cheerful representation of China, it pitched the argument that European and Chinese cultures derived from a common origin of belief in God. 'The Chinese, from the beginning of their origin to the time of Confucius,' Couplet explained in an abridged popularisation of the book, 'paid adoration only to the Creator of the universe.' Not knowing the name of God, they called Him Xam Ti, but they knew that He existed and even built temples to Him. The Jesuits reasoned that Confucius was to be revered, for without him the Chinese would have drifted into polytheism and been for ever estranged from their primitive Christianity. The Christian religion was not a foreign import that had to be imposed on the Chinese, then; it was merely something they needed to be reminded of and brought back to. Accommodation, not conquest, was the path that missionaries should follow.

Confucius caused a lot of excitement when it appeared in May 1687. Couplet in his introduction reminded readers that the book was not intended simply to satisfy idle curiosity, yet the buzz around the book, like the buzz around Michael Shen, was surely not unintended. The book was important enough for Jesuits to present James II with a copy, and for Hyde to have acquired a copy for the Bodleian as soon as it came out.*

Once the book was in production, Michael was sent to London to meet Jesuits there and, should the occasion arise, to charm the king. The strategy worked. The king was indeed charmed, to the point of commissioning the royal portrait painter, Godfrey Kneller, to do Michael's portrait (Fig. 9). It is a fine work. Kneller could have dashed off a generic Chinese face, and there is a French engraving of Michael that does only that. Instead, the painter has lit the face carefully to suggest an

---

* The copy of *Confucius Sinarum Philosophus* that James noted on his visit to the Bodleian was later sold as a duplicate. The present copy was purchased at a book auction in Paris in 1825. I am grateful to David Helliwell and Sarah Wheale of the Bodleian for tracking down this curious fact.

individuality of person and devotion. Its customary title is *The Chinese Convert*, and it certainly stages that idea. Michael declares his piety by raising a large crucifix bearing a crucified Christ in his left hand and gestures to it with his right. He has turned his head well to his right and is looking up towards heaven. Behind him to his left there is a table draped in an Oriental carpet, and on that carpet rests a richly bound tome. On top of that lies what looks to be a sheaf of manuscript. Conventionally it has been assumed that this is a Bible, the pole star for the new convert. But the tome I believe I see in Kneller's painting is not the Bible but *Confucius Sinarum Philosophus*. If I am right, it lies there as a quiet reminder of the accommodationist position that argued for bringing Chinese to Christianity without forcing them to abandon their own culture. Beside it stands Michael Shen as living proof of the argument.

But there may be something else going on, if we allow Michael to have had a say in the props that went into his painting. *Confucius* was a major undertaking that demanded a thorough knowledge of Chinese language, philosophy and culture. Couplet was a man of great talent, skilled in the Chinese language, but he would have needed a native informant to complete the work of fact-checking, filling in missing details and going over the Latin translation against the Chinese original. Michael was his only living link to the culture he was interpreting to Europe. Couplet needed him, and after the social rounds were over, he must have put him to work on this project. Couplet published another book later the same year, the exemplary life of Candida Xu, a convert in Shanghai and granddaughter of the Catholic Grand Secretary, Paolo Xu. Michael is not named in that book, but he must have had a hand in it. The clue lies in a file of manuscripts collected by the Irish physician Hans Sloane and donated to the British Museum. The file, labelled 'Sloane 853a', contains Thomas Hyde's notes on Chinese matters. In this file there is a slip of paper on which Michael has written Candida's name and ancestry in Latin and Chinese. This may not be conclusive proof that he helped Couplet produce the book, but it suggests that he was more than an ornament to garner the Jesuits' social attention and attract patronage. If my hunch is correct and it is indeed a copy of *Confucius* in the Kneller portrait, it may have been Michael's own touch: a subtle way of reminding the world that the great book was his too.

While Michael was meeting James II and sitting for Kneller, Thomas Hyde was waiting for him in Oxford. The two had been in correspondence since the end of 1686, while Michael was still in Paris working with Couplet. As Keeper of the Bodleian Library, Hyde had a problem that Michael could solve if he came to Oxford. To understand how he got into his problem, we need to go back to 1659, the year the Selden bequest arrived at the Bodleian, and the year Hyde became Under-Keeper. (He would replace Thomas Barlow as Keeper six years later.) It was a truth universally acknowledged in the seventeenth century that a good library required a catalogue, so it was understood from the very beginning of Hyde's tenure there that he had to create one for the Bodleian. Hyde was less than enthusiastic. He loved nothing better than the tedious challenge of deciphering manuscripts in obscure Oriental languages, but for the tedious work of cataloguing he was unsuited. The job got done, though not without much complaining and cajoling of junior associates including Anthony Wood, who disliked him for it.

Wood's diary is our source for most of what we know of Hyde the working librarian, and he doesn't come off well there – although, to be fair, almost no one does. The two started off on the wrong foot. Wood asked the vice-chancellor of the university for unrestricted access to manuscripts in the Bodleian, in return for which he would help Hyde by fetching books and manuscripts that readers called for. Hyde turned this arrangement to his advantage by going to the vice-chancellor behind Wood's back and making access conditional on Wood helping him catalogue the manuscripts. 'Mr. Hide did not carry himself like a gent', is Wood's comment in the diary. The two eventually worked out some sort of accommodation, for Wood was still there in 1687, when James II came for breakfast. But that didn't stop Wood from scribbling in his diary that same year that Hyde's wife was a whore, a Catholic and 'now a madwoman'. Wood is not an impeccable source, but whatever her condition, Hyde's wife died later that year.

One thing that set Hyde apart from Wood was his claim to fluency in several Oriental languages, a claim without which he wouldn't have got the position at the Bodleian. This was because of the collecting practices of the Bodleian. One of the working principles of the library right from the beginning was that it should collect materials in all languages, and in they had come, many by purchase and some by donation, many as renowned texts of scholarly value and some as curiosities. When Hyde arrived in Oxford in 1657, the books and manuscripts in Oriental languages were starting to pile up. He had actually been hired to teach Hebrew, but he was hopeless at teaching. As he later acknowledged in the letter announcing his retirement, the auditors at his lectures were 'scarce and practitioners more scarce'. He soon angled his way over to the Bodleian, where his skills – not just in Hebrew but in Arabic and Persian as well – were by this time much needed. Hyde was the man for the job, and he was on site.

Knowledge of Oriental languages was not a common skill at the time, but, thanks to the inspiration of scholars such as John Selden, it was a skill in some demand. The biggest project in mid-seventeenth-century Europe that required this knowledge was the production of Polyglot Bibles – that is, Bibles in multiple ancient languages. The French were the first to publish a Polyglot Bible, in 1645, an unwieldy but impressive production in ten lavishly bound volumes. The Bodleian acquired its copy in 1649. The English followed suit, although the project came about in a rather circuitous way. The Christian divine who took it on, Brian Walton, did so because he could gain no other employment. Denounced by his parishioners in 1641 for being a royalist, he fled to Oxford, then King Charles's stronghold against Parliament's forces. After the collapse of the royalist camp in 1646, Walton returned to London and lived unemployed in his father-in-law's house. Sidelined by politics and out of work, he decided to compile a Polyglot Bible. When it was published, in 1657, it consisted of the authoritative Latin text accompanied by the same text in Hebrew, Greek, Syriac, Chaldean (referred to as Aramaic today), Samaritan and Arabic, each with its Latin interpretation, all laid out across each pair of facing pages. Its two main promoters were none other than John Selden and the man who taught him Arabic, Archbishop James Ussher. They co-authored an appeal to prospective investors, complimenting Walton's 'Method and order wherein the severall Languages are digested' and declaring that the edition 'will much tend to the Glory of God & the publique honour of our Nation'.

This is where Thomas Hyde comes into the story. Hyde's Arabic professor at Cambridge, whose only (failed) attempt at Arabic scholarship was to write a refutation of the Koran, died in 1653, leaving a great deal of work undone. Walton decided to bring in this professor's student to fill the gap. Hyde thus became the youngest of the Orientalists to take part in the production of the Polyglot Bible. Walton singles him out in the Preface as 'a youth of the highest promise, who has made great progress, beyond his years, in the Oriental Tongues'. It was a shining start to Hyde's career, and his ticket to Oxford. The ticket to Oxford turned into a transfer to the Bodleian, although Hyde still yearned for a professorship that would honour his talents and furnish him with a better stipend. That eminence eluded him for decades, in part because he published no research beyond a series of studies of Asian board games. He did finally publish his research on Persian religion in 1700, three years before he died. *Historia Religionis Veterum Persarum* attracted attention among scholars on the Continent but earned so little acclaim in England that it was said he burned part of the print run to boil his tea. He did finally obtain a professorship – of Arabic – in his final decade at Oxford, although he was passed over for the Hebrew professorship until just a few years before he died, when the incumbent was fired for refusing to swear allegiance to King William and the honour fell to him.

Despite his naturally lethargic disposition, Hyde loved to learn languages – the more the better – and took every opportunity to talk to a native speaker when he was learning a new one. To the east lay the richest trove of unlearned languages. To Hyde, Asia unrolled as one unconquered linguistic peak after another, each more remote and more enticing than the one before it. He wanted to climb every one. As late as 1700 he was still actively trying to acquire documents in Sanskrit and Telugu and was casting about for the 'Tartar Alphabet of Kithay', the script of the Manchus, the Tungusic people who conquered the Ming dynasty in 1644. Gerald Toomer, the authority on the history of Oriental studies in this period, is generous in his praise of Hyde, noting that he 'had an extraordinarily wide interest in oriental tongues, and was concerned to master the spoken as well as the written languages by conversing with native speakers'. Hyde's appetite for this knowledge was so strong that late in life he could come up with a long list of the works he still dreamed of publishing: a translation of the New Testament in Malay (he supervised a printing of the Gospels in Malay in 1677), a Hebrew edition of Maimonides with Arabic and Latin transcriptions, a Hebrew lexicon with Arabic and Persian equivalents, and an Arabic edition of the Pentateuch with Latin glosses.

And then there was Chinese, a language that beckoned to him from what must have seemed an impossible distance. There were Chinese books in the Bodleian that needed identifying and cataloguing, certainly, and that may have been the excuse to bring Michael to Oxford. But the invitation was also the best way he could bring himself into contact with the language that fascinated him for its own sake.

The challenge of learning it was thought to be formidable. 'The Chinese language', one Jesuit author assured his readers,

> has no analogy whatsoever with any other language in use
> throughout the world. Nothing in common: neither the sound of its
> words, nor the pronunciation of its phrases, nor the arrangement
> of ideas. Everything is mysterious in this language: one can learn
> terms in two hours, yet it may take several years of study to be
> able to speak them. One can learn to read all Chinese books, and
> understand them well, without comprehending anything if someone
> else recites them. A scholar will be able to compose essays rich in
> elegance and polite phrasing, yet the same scholar will not always
> know enough to explain himself in ordinary conversation.

Worse still, 'the same words often signify opposite things, such that when two people pronounce them, what will be a compliment coming from the mouth of one will be atrocious insults from the mouth of the other.' The language could still be learned; indeed, it could become 'fertile, abundant, and harmonious in the mouth or under the brush of those who have applied themselves to its study'. But it wouldn't be easy. From there it was only a short step to the declaration of George Bonham, a nineteenth-century governor of Hong Kong, that it was unwise to study the language, as it 'warps the mind and imbues it with a defective perception of the common things of real life'.

All of this is complete nonsense, of course, but it passed as sensible judgement at the time. Hyde in his passion for Asian languages was undeterred. And so he brought his Chinese guest in Oxford in June

1687. Michael stayed for roughly six weeks, although the exact dates are not recorded anywhere. The only documentation of Michael's stint as a Bodleian cataloguer is an entry in the library accounts for 1686–7 regarding the sum of £6: 'Paid the Chinese for making catalogues to the China Bookes, for his expences and Lodging.'

Michael spent part of his time engaged in this task, working his way through the several dozen Chinese books and manuscripts that had accumulated in the library since 1604, noting on their covers what they were. His annotations in a clear hand, both in Chinese and in Latin, can still be seen on these books today. He also drew up lists of what was in the collection, drafts of which can be found in Hyde's file in the British Library. But he also spent much of his time with Hyde, to judge from other notes preserved in the file. Hyde wanted to learn the rudiments of the language, and Michael taught him. On one scrap of paper he has written (in Chinese characters) *xiyang* ('the West') and *zhongguo* ('China'), one above the other, then penned beside them a four-character catechism for each. In the West, 'from how it sounds you determine how it is written'; in China, 'from how it is written you determine how it sounds.' It is a neat summary of the difference between alphabetic and ideographic languages. It certainly suited the mind of Hyde, who was always prowling for keys such as this to make learning languages easier.

The Chinese who emerges from these scraps of paper is a figure in fragments. He identifies himself as a native of Nanjing and 'junior in age' to Hyde, whom he addressed with the honorific title of *laoye daren*, 'respected elder'. Two letters he sent to Hyde are addressed: 'For mr. Dr. Hyde chief Library Keeper of the University of Oxford at Oxon' and 'For Dr. Hyde chief Library Keeper of the University of Oxford to be found at Queens College Oxon'. Michael seems to have mastered enough English to address a letter. It is also worth noting that the original red seals on two letters have been preserved. The image pressed into the wax is of a two-masted ship at sea. Michael Shen, sailor of the world.

The Englishman lurking in these notes has a fascination for all things Chinese, although there is a random quality about the subjects he inquires into. Hyde has an interest in religion, getting Michael to run through the various names for Buddhist temples and provide a short description of Buddhist worship. He also takes notes on how paper is made from bamboo, silk and 'the inner Rind of a tree'. Michael has written the character for 'cotton', but Hyde seems not to have figured out what it meant. He also writes a short note on the Chinese gun, following Michael's account. Their firearms were fired not with a steel hammer on flint – a technique that European gunsmiths had developed earlier in the century – but with a fuse, and not a fuse held in a matchlock (a European innovation from late in the sixteenth century) but in the hand. Because of this primitive arrangement, Michael explains to Hyde, when Chinese gunners 'fire' their guns, they 'turn their faces back for fear'. Not a good pose from which to hit a target.

The most consistent language-learning Hyde did was numbers. This is always a good place to start, given the complete lack of ambiguity of numbers. On one sheet he has written in a column the numbers 1 to 13, then 20 to 40 by tens, then 100 and either 1,000 or 10,000 – it is hard to tell which number he has written, since he has crossed one out and written in the other. His confusion is not surprising, for numbers are not just numbers: they operate within systems. In England the system for higher numbers is to count them in thousands, whereas in Chinese they are counted in tens of thousands. Once recognised, though, systems are refreshingly context-free, even content-free. To Europeans they were mechanisms that logic could unlock once you had the key. The key to Chinese – spoken of by Hyde and many others as the *clavis sinica* – was what every hopeful Orientalist longed to find so that he could decipher Chinese without having to go to China and spend years learning the language, which would be the fate of mere mortals like me. It is possible to learn Chinese, but it cannot be done by picking up a shiny key and unlocking the language with a twist of the wrist.

Where we can watch Michael and Hyde working together is on the Selden map. The original writing on the map is entirely Chinese, but beside many of the labels are ghostly translations and annotations in spidery European letters. The writing is so minuscule, the ink so faded and the paper so worn that it is easy to miss what they have written, even if you get the chance to see the map in real life. But they are there. Michael's distinctive hand always comes first, showing how to pronounce each character. His romanisations are then followed by Latin translations in Hyde's rather more cramped hand, sometimes character by character, sometimes word for word. To anyone who has applied for a library card at the Bodleian, the scribblings come as a surprise. The memorable moment during registration is having to recite a solemn promise, which begins (the original in Latin, of course): 'I hereby undertake not to remove from the Library, nor to mark, deface, or injure in any way, any volume, document or other object belonging to it or in its custody.' (The pledge goes on to enjoin readers not to kindle fires inside the library, a hold-over from the centuries without central heating.) One supposes librarians weren't so squeamish in earlier days, besides which Hyde was in charge. For studying the history of the map, however, this lapse is a godsend, because it lets us watch Michael and Hyde at work.

The romanisations are correct, if unconventional, but there were as yet no conventions for writing out Chinese in roman letters (for that matter, English spelling hadn't even been standardised). Michael's romanisations betray a mix of the official dialect – what today we call Mandarin, that word being a Portuguese derivation from the Sanskrit word *mandarim* ('official') – and his native dialect as someone who grew up in Nanjing. One difference between them is the *-ng* ending in Mandarin, which becomes *-m* in Nanjing dialect; another is the use of *v* in Nanjing dialect for *w* in Mandarin. Thus the word for 'king' (*wang*) is sometimes *vam* in the notes but *wang* on the map, although one half-corrected *wan* makes an appearance. Michael is more casual about these distinctions in the notes, but on the map he does his best to stay within Mandarin pronunciation.

Occasionally the two men go beyond romanising and translating. For example, there are two labels to the right-hand edge of the map that Michael romanises as *hua gin chi* and *hung mao chi* (we would romanise these today as *hua ren ju* and *hong mao ju*). That is all either man writes on this part of the map, but the margin has more. Most of the original margin around the map has flaked away to nothing, but a fragment of the margin beside these labels has survived, and there we can see fragments of notes that Michael and Hyde made. Michael has repeated the three syllables *hung mao chi*, to which Hyde has added translations, although the only word still legible is 'capillus', meaning 'hair', for *mao*. Hyde has written *hung=mao* again below this, connecting the two syllables, as Michael never does, with his characteristic double hyphen (=), and he has added a translation: 'Hollanders'. The term *hung=mao* means 'red hair', and was the name Chinese gave the Dutch because of the shocking (to Chinese eyes) colour of their hair and beards. The label *hung mao chi* means 'Where Red Hairs live' and refers to a Dutch outpost in the Spice Islands. Interesting for their own sake, these marginal scribbles hint that perhaps the entire map was surrounded by such annotations. Analysis shows that the paper used for the border was a European one manufactured from hemp and was therefore added later, presumably in England. It may not have been intended to provide Hyde with a sidebar on which he could make notes, but this appears to have been how he used it. Too bad the border strips have turned to dust. It would be interesting to see what else caught his attention.

The annotations could suggest that Hyde pored over the map out of a keen interest in geographical knowledge. This first impression seemed to find confirmation when I found a map in Hyde's file in the British Library that Michael drew freehand. It is a map showing the Great Wall extending from the Gate of Eastward Domination at the coast, which is not on the Selden map, to Jade Gate Pass in the west, which is, and beyond that to points as far west as Turfan and Samarkand. To the north Michael has filled in some two dozen rivers and mountains across a terrain extending across Siberia to *Bei Hai*, 'the North Sea' – Mare Septentrionalis to Hyde, the Arctic Ocean to us. I took this to signify that Hyde had a keen interest in the physical geography of Siberia, perhaps because this was the zone from which the Manchus had invaded in 1644, after the Selden map was drawn. My hypothesis collapsed when David Helliwell showed me a pair of extraordinary Chinese printed maps, also in the Bodleian, also from the seventeenth century and also annotated by Michael. Michael had simply hand-copied the northern section of one of these maps so that he and Hyde could make annotations without marking up the original.

My sense is that Hyde was struggling to understand not China as a place but Chinese as a language. His notes on the map and the word lists in his file of notes at the British Library suggest someone trying to build up a vocabulary in a new language. But there is nothing whatsoever about how the language works grammatically. To put this pursuit a little starkly, he was collecting words, not language. But what else could he have done in barely six weeks?

A nineteenth-century scholar who knew no Chinese credited Hyde with having 'made great progress in the language of China', but the evidence does not persuade me. That he tried at all is to his credit. He even used it in his publications. Chinese characters appear in his book on Persian religion, as they do in his book on Oriental board games, which includes several pages of characters in Michael's calligraphy. One character is printed upside down, but that would be the fault of the printer. On one page where Hyde demonstrates how Chinese characters are formed, the characters look as though they had been written by a beginner copying someone else's handwriting: the student copying the teacher, Hyde copying Michael.

In his own mind Hyde believed that he had enough of a grasp of the language to go into print with it. The most curious sheet of paper in the British Library file is the one on which Hyde has mocked up a notice or maybe even the title-page of a book he would like to publish. The long-winded title begins: 'Adversaria Chinensia à scripto et ore nativi Chinensis excepta, in quibus sunt Decalogus, symbolum Apostolicum, Oratio Dominica, Ave Maria, Grammaticaliae et Formulae.' We might translate this as: 'Observations Taken from the Writing and Speech of a Native Chinese, in which are Rendered the Ten Commandments, the Apostles' Creed, the Lord's Prayer and Hail Mary, with Grammar and Phrases.' Some, but not all, of the texts the title-page promises can be found among Hyde's papers: not enough to complete the planned book. As I was writing this book, however, Frances Wood, an old friend from student days in China who now heads the Asian section of the British Library, wrote to say that more documents in Michael Shen's hand had just surfaced in another file in the Sloane collection. She sent me photocopies. In addition to half a dozen Buddhist and Daoist prayers, the new file included fair copies of the Lord's Prayer, the Apostles' Creed and the Ten Commandments, written in large Chinese characters with subscript romanisations. The Chinese is in Michael's hand, the Latin in Hyde's. These are the rest of the manuscript copies of the book Hyde hoped to publish, exactly as promised on his title-page. Clearly a publishing project was under way with Michael's help; just as clearly, it never came to fruition.

When it was time for Michael to leave Oxford, Hyde sent him to

London with an introduction to the eminent scientist and philosopher Robert Boyle. He hoped that Boyle might be able to steer funds his way to support publication, and that meeting Michael would ignite Boyle's enthusiasm for the project. Boyle was intrigued by Michael, but what would have been the first English book on Chinese never appeared.

———————————

Thomas Hyde cherished Michael's memory for the rest of his life. Calling him 'my Chinese' in a letter, he wrote: 'Michael Shun Fo-Çung (for that was his name) was bred a Schollar in all the Learning of their country, read all their Books readily, and was of great honesty and sincerity, and fit to be relyed upon in every thing.' The praise may have exceeded the achievement, but it was to remember a friend with whom he had lost contact. He would recall him just as fondly in print in his magnum opus on Persian religion, as 'my Chinese friend Dr Michael Shin-Fo Çungh of Nanjing, who today lives in Nanjing', not knowing that Michael was dead.

Shortly after Michael sent his last letter to Hyde, on 29 December 1687, he and his mentor Philippe Couplet sailed from London to Lisbon. Their intention was to return immediately to China, but the politics of French Jesuits travelling in Portuguese ships was sufficiently tangled and tense that the two ended up stranded in Lisbon. Michael spent three years there, studying to take his first vows towards becoming the first Chinese Jesuit priest. He was finally cleared to leave Portugal, but he died at sea, probably of dysentery, somewhere between the Cape of Good Hope and Mozambique. Two years later Couplet would take the same return journey and meet death on the same ocean.

Hyde was left behind not just by his Chinese informant and, yes, friend. He was left behind by an age that was rapidly losing its fascination for Oriental knowledge. During John Selden's time Oriental studies had promised to hold the keys to all sorts of secret knowledge of human history and human institutions. The world was still opening up in the first half of the seventeenth century, and it was clear to the brightest scholars that the richest troves of texts and traditions lay to the east. Half a century later, this conviction faltered. By the time Hyde was negotiating his retirement after completing what he regarded as his masterwork on Persian religion, no one much cared any more. Had Selden written

Historia Religionis Veterum Persarum, readers would have been snatching it from the shelves as soon as it appeared. The problem wasn't just that the great Selden was not the author. It was that the time for believing that such work must point the way forward to new and startling discoveries had passed. Things Oriental were becoming things ornamental, matter for amusing conversation with the king, perhaps, but not the stuff of serious discussion with one's intellectual peers. Selden could see himself standing at a new dawn of knowledge creation, likening Oriental languages to Galileo's telescope: both would yield discoveries of which earlier generations had not even dreamed. What the historian Nicholas Dew has termed 'Baroque Orientalism' held fewer such promises. Not too long thereafter the study of Asia would slide into what Claire Gallien has dubbed 'pseudo-Orientalism', a field not of scholarship and critical insight but of fantasy and amusement: all that Coleridgian nonsense about 'stately pleasure-domes' and 'caverns measureless to man'.

The young Hyde had caught the tail end of the Orientalist wave when he was brought in to work on the Polyglot Bible in the 1650s. The mature Hyde had used his Orientalist knowledge to compile the first catalogue of the Bodleian Library in the 1670s. The elder Hyde couldn't sell his magnum opus, and flogged part of his collection of Persian and Arabic manuscripts to the library that employed him for cash. Compared with what Selden had done and what he had donated fifty years earlier, this was a meagre showing. The man whom Brian Walton had praised as a 'youth of the highest promise' hadn't lived up to expectations – of others or of himself. It wasn't his fault. The world had changed around him.

———————

With the office of Keeper came the privilege of having a portrait painted for the benefit of posterity. The accumulated portraiture of Keepers hangs today in the old School of Astronomy and Rhetoric – now the Bodleian's gift shop. Hyde's has been given the spot above the cash register (Fig. 7). The painting is standard for the style prevailing at the end of the seventeenth century. It is not a great work of art, but it is not without interest, especially for us. It shows its subject from the waist up against a dark background, his body turned to the left. Dressed as a divine, the subject gazes impassively at the viewer from under the then recently mandatory fashion accessory of a wig. Everything but the face and hands is generic and could have been painted by an apprentice without even requiring the sitter to be present. The face suggests to me someone who has not quite attained the dignity he desired and who will stare down anyone who points this out to him; also someone who sees no reason to give much of himself away. His hands look out of proportion to the body, but there is something there of interest. The left hangs impassively at his side, but the right is slightly raised. This is where we can catch sight of a curious detail that the sitter has consciously staged for his viewers' benefit, for in his right hand he holds a sheaf of rolled-up paper. Something is written on it. If you stand at the cash desk and look up, it is hard to read what it says. When the woman at the cash register noticed me peering up at the portrait, she brought over a small step-ladder so that I could get a better look.

Even standing at the counter, you can probably recognise the writing as Chinese characters, two appearing below his fingers and three above. The two below are *jin* ('metal' or 'gold') and a character that might be *gu* ('ancient') or *she* ('tongue'), although an errant dot on the left side makes it impossible to be certain what was intended. Of the three characters above his fingers, two are very clear: *gu* ('ancient') and *li* ('principle', also the unit for a distance of a third of a mile). The third character, which curves around the sheaf of paper, could be *zhou* ('prefecture'), but that's a guess.

Once I was up the ladder, I could better assess the calligraphy. I was looking at practice characters written by a student in his first term of a class in introductory Chinese. That is, he knew which way was up and which was down and had a fair sense of how a character is formed. He has drawn *li* by making the line second from the bottom longer than the bottom line, a fundamental error a teacher would immediately correct. He has also simplified the third- and second-to-last strokes of *jin* into a single line, something he might have seen a veteran calligrapher do, but the way he has done it does not look quite right. All of which is to say that Michael was no longer around when Hyde himself painted these characters into his official portrait.

One wonders why Hyde chose to be portrayed with Chinese characters. He could have posed more successfully with Hebrew, Arabic or

Persian script. These languages he knew well. Why did he want Chinese? Was it because any (albeit small) number of people in Oxford could read Hebrew, Arabic or Persian, whereas no one could read Chinese? Was he showing that, yes, he had mastered the most elusive Oriental language in the Bodleian catalogue? Was he seeking to elevate himself above the already exclusive club of Orientalists by declaring that he alone had mastered Chinese? Or was learning some Chinese simply the most exciting study he ever did?

These suspicions pushed another question to the forefront as I stared up at his portrait. Hyde never learned that Michael Shen succumbed to sickness on board ship between the Cape of Good Hope and Mozambique, but surely he should have imagined that the time would come when people other than Michael who actually knew Chinese would look at the portrait and unmask this gentle fraud. Perhaps I am being unfair. Who of us can look into the future and guess with any confidence what will become common knowledge and what will sink into obscurity? Where Hyde stood at the end of the seventeenth century, there were no Chinese at hand and none on the way. He stood alone, a sentinel at a lonely frontier post.

Just before retiring as Keeper in 1701, Hyde wrote to the Archbishop of Canterbury to offer advice regarding his replacement. The new Keeper must 'have the advantage of the Eastern languages,' Hyde stressed, 'otherwise, he will be much in the darke'. It was sensible advice, but the tide of scholarly fashion was already flowing in a different direction. The wonder-seekers were looking elsewhere by then; they still are, for that matter. No scholar of Oriental languages will ever again head that library. As for deciphering Chinese documents, there are specialists for that sort of thing. Like me.

We're finished in the library. As we leave the stacks and head to sea, we must also move back in time to the maritime trading world that the Selden map captures, back before John Selden ever picked up the thread that started at his meeting with King James.

4

John Saris and the China Captain

By the third week of December 1614 the publicity had become unbearable. Thomas Smythe knew that he, as Governor of the East India Company (EIC), had to do something to kill the rumours swirling around London. John Saris had arrived on the *Clove* loaded to the gunwales with rich pickings from Asia, but he had the audacity to anchor at the south-coast port of Plymouth to transact private business before bringing the ship up the Thames to London. The longer he dallied at Plymouth – which his commission expressly forbade him to do – the louder commercial London buzzed with nasty gossip about how the Company was being robbed blind by its own servants. This was not to be borne. The EIC had existed for only fourteen years. It was a canny blend of private capital and royal charter created by the grace of the late Elizabeth I for the benefit of the merchants who supported her. James I had replaced her, and so now the Company's survival as a monopoly depended on keeping the present king happy and potential competitors out of the Asian trade.

The question of when Saris would actually get the *Clove* to London was the most recent in a series of headaches that Smythe had had to deal with to keep the EIC above political and financial water. Recently he had had to cave in to the Lord Treasurer, who had insisted that the Company open a shop in the New Exchange, which he had authorised to be built as a new commercial hub for the city (and for which the opening night's entertainment in 1609 was penned by none other than Ben Jonson). Smythe was supposed to furnish it with Chinese paper, fans, ink boxes and porcelain, but such goods were expensive and hard to come by. Why waste money displaying wares to idle gawkers? And would Saris's cargo make good the £300 it had cost him personally to set up the shop? All Saris had forwarded so far was a Malayan dagger worth £6 or £7. There were, in addition, rumours of Saris beating his officers and starving his crew, but everyone whined about discipline at the end of a successful voyage when the prospects of a pay-off were high. And there was also James I to be kept happy. Smythe had been reckless enough to inform the king that he could expect a lavish present from the emperor of Japan (in fact, it was from the far more powerful shogun, but that title meant nothing in 1614), but when would that be delivered?

Smythe had no trouble deflecting the slanders of 'malicious and scandalous tongues' against John Saris at the general meeting of members of the EIC on 6 December. As its Governor, he had constantly to answer stockholders' charges against his employees, and he was good at it. His standard trick was to remind them that their enemies would be only too happy to watch over the destruction of the Company. Solidarity within the Company was the only protection against this sort of ganging up from without. The stockholders 'should not be enymies to themselves by condempninge this there [their] comander', as the minutes of the meeting have him saying. Fortunately, he was able to report that the auditors were already estimating that Saris had tripled the capital put into the voyage. Things would turn out well as long as everyone stayed calm.

The allegations about problems inside the Company were nothing, however, compared with the looming pornography scandal. The Company had the good sense not to inquire too closely into what men on ships got up to, yet the charge that Saris had brought Japanese pornography into England would be an easy excuse to spark the malicious into alleging that the Company lacked a moral rudder and should be stripped of its monopoly on Asian trade. Big business had to appear to be righteous business, or someone outside the company would want to tear it down and keep a piece of the business for themselves. Ten days later Smythe had to report to the Court of Committees – the Company's board of directors – regarding 'some imputacions and aspersions beinge cast upon Capt'ne Saris for certain lascivious bookes and pictures brought home by him and divulged'.

Bringing erotica home from Japan was one thing; showing it around was something else entirely. The directors felt that Saris should have shown better discretion than to display pictures of men with prodigious genitalia copulating with compliant female partners – certainly the most spectacular erotica that anyone in England had ever seen. Saris's lack of judgement had unleashed 'a greate scandall upon this Companye', which the directors regarded as 'unbeseeminge their gravitie to permitte'. They might have put their gravity aside and let the scandal die its own quiet death, had word of the pictures not got out. But it did. Notoriety forced Smythe to act. Assuring his directors that he disliked the pictures as much as they did, he told them that he knew the offending materials were in Saris's house, and that he would 'gett them out of his haundes yf possiblie he could'. His recommendation to the Court of Committees was that the books and pictures 'bee burnt or otherwise disposed of as the Company should thinke fit'.

As well as being Governor of the East India Company, Thomas Smythe was also Treasurer of the Virginia Company, which was setting up plantations in the New World. In addition, a decade earlier he had travelled to Russia to negotiate terms of trade for the Muscovy Company. Each of these companies enjoyed the protection of a royal charter, giving it a monopoly on English trade in and out of Asia, North America and Russia respectively. Smythe had major stakes in all three corporations: there was no more powerful businessman in London. In the moral vacuum of business, the appearance of 'gravitie' was everything. Smythe had to be seen to do something. Whether or not he was actually offended, he was canny enough to anticipate how accusations of immorality could hurt his business interests.

Three weeks passed before Smythe reported back to the Court of Committees on the matter. As he told his directors, because of the 'greate speeches' made against the Company and against him personally on the Exchange, he had confiscated Saris's books and pictures and 'shut them up ever since'. Perhaps he had hoped that would be sufficient to still the little storm, but it hadn't been. The only course now was to destroy the offending pictures in public view, so that anyone who had been 'honestlie affected', as he put it, would know that action had been taken. It would be an unambiguous demonstration that 'such wicked spectacles are not fostered and mayntayned by any of this Companie', he declared. 'And thereupon in open presence [he] put them into the fire, where they continued till they were burnt and turnd into smoke.'

Thus it was that the first Japanese erotic prints to find their way to England were consigned to oblivion. Other tokens from this early moment of contact survived censure, although not all have withstood the wear and tear of time. The two suits of Japanese armour that Shogun Tokugawa Ieyasu sent to James I were sent to the Tower and are still in the Royal Armouries, but the set of ten large folding screens have long since disappeared. Saris describes them as 'large pictures to hang a chamber with', a turn of phrase from a culture in which pictures were hung rather than stood on the floor, as they could be in Japan. Not only is the gift gone; so is James's reaction. But we do know Thomas Smythe's, which is that they were sub-standard. As the shogun never saw the screens he ordered, we have no way of knowing his intentions. Presumably he meant to send good work, and his minions took a cut and delivered second-rate goods to Saris. Whatever was the case, Smythe decided to substitute several of the Company's own screens for the ones Tokugawa sent. Everything must be done to ensure that his political master receive the best of what the Company received, and thus the best impression of Tokugawa's regard. After all, this was a diplomatic gift: neither aesthetics nor authenticity mattered. It just had to look good.

The appeal of pictures at both ends of the England–Japan trade was something the EIC soon learned to anticipate. We see this in the account books of the factory – the term used at the time for a foreign commercial post – which the Company set up in the port of Hirado, west of today's Nagasaki. These inventories list dozens of paintings and prints in stock. Some were to be offered for sale, others were given as gifts; yet others had to be written off as damaged goods after a voyage. The commonest subject of oil paintings at the factory was Venus, with or without Adonis, Bacchus or Cupid. Pretty women, the English assumed, would always be a welcome sight. The factory inventory also included several copies of the king's portrait, plus such standard fare as the four seasons, the five senses and landscapes by the dozen. The most expensive painting listed in the factory accounts of June 1616 was an oil portrait of none other than Governor Thomas Smythe. At the stunning valuation of £12, the head of the Company cost four times the best of the eight Venuses in the inventory.

Saris believed there was money to be made in exporting English art to Japan. While still lingering in Plymouth, he sent the London office a list of English products that would sell well in Japan. The first part is a detailed account of fabrics – these proved not to be popular. Then there follows a list of other items, at the top of which is: 'Pictures, paynted, som[e] lascivious, others of stories of wars by sea and land, the larger the better.' When he repeated this advice before the Court of Committees later that winter, after the pornography scandal had been put to bed, he commended battle scenes and avoided any mention of 'lascivious' pictures. But we know where his taste lay. In his ship's journal he records that in his own cabin he had a 'picture of Venus hung, verye lasiviously sett out and in a great frame'. He mentions this in the context of reporting how some Christian Japanese women fell to their knees when they entered his cabin and saw it. They thought it was an image of Mary.

In the end, there wasn't much of a market in Japan for European art. Part of the problem lay simply with the difficulties of keeping artwork clean and dry in a ship's hold. Two years after Saris's return, the merchant he left behind in Japan to manage the Company factory, Richard

Cocks, wrote back to London telling the Company not to bother sending any more oil paintings. Not only were they vulnerable to the shipboard damp, but their foreign aesthetic placed them outside what Japanese would dignify as art and therefore not worth paying money for. Rather than paintings, send prints, he advised. Paintings 'they esteem not, but had rather have printed black paper with shipps, horses, men, battells, burds or suchlike trifles'. The inventory of visual images in the Hirado factory reveals that maps of Britain outnumbered paintings and prints by two to one, but Cocks does not mention whether he thought these could sell.

———————————

Thomas Smythe had appointed John Saris three years earlier to command the eighth voyage of the East India Company to Asia. He left England in April 1611 and after trading along the coast of India reached Bantam in October 1612. He soon set sail for the Spice Islands to prospect for spices. While his first obligation was to make money for the Company, he was sent with a political purpose as well, which was to challenge the recent Dutch effort to displace Spanish dominance of the spice trade in the Moluccas – the Spice Islands. These tropical volcanic islands are scattered between Sulawesi and New Guinea towards the eastern end of today's Indonesia. The ecology of the wettest northern islands in this archipelago – Ternate, Tidore and Matyan – was perfect for growing nutmeg, cloves and mace, spices that fetched their weight in gold back in Europe. The trade was an old one, controlled by local chiefs and handled by Chinese and Muslim traders before Europeans arrived.

John Saris was there in effect to challenge the VOC position laid out by de Groot in *The Free Sea*. If the seas of East Asia were as free as de Groot said they were, the English should be able to trade there. This wasn't how the Dutch saw the situation. Having chewed off parts of the Portuguese and the Spanish empires, they had no intention of letting the English in on the spice trade. Saris was there to test the proposition. The Dutch were completely unwilling to let the English into the trade. Every time Saris approached a local ruler to set up a trade deal, he found himself blocked by the Dutch. Even small sales were scuppered. Whenever a prospective seller came forward with cloves to sell, Dutchmen who were there before him used 'a mixture of coaxing and intimidation', as the historian Martine van Ittersum has phrased it, to prevent the English from doing any business.

When Saris reached Matyan, in mid-March, a local headman came aboard the *Clove* to discuss trade, but two Dutchmen accompanied him to make sure the man didn't sell the Englishmen anything and to threaten the pilot who had guided Saris to the island. The Dutch had taken control of Matyan five years earlier, and they insisted the English had no right to be there. Having conquered by the sword, the spices of Matyan were their due by right of conquest. When Saris told Richard Cocks, his third-in-command, to kick them off his ship, the two Dutchmen threatened that they would kill anyone caught selling cloves to the English. Saris replied stoutly that he would trade 'with anye man which would deale with me'. In practice this proved difficult to arrange. The locals tried bringing cargo aboard the English ship that night but were surprised and surrounded by heavily armed Dutchmen. Saris sent Cocks ashore the next day in hope of reviving the deal, but Adriaan Martens Blocq, the Dutch commander, 'came to Mr. Cocks, and willed him to tell me that I sent my people no more ashoare in the night, for yf I did he would kill them. Mr. Cocks answered him with laufture, and so lefte him.' Laughter was a bold response from Cocks, but it was not enough to loosen the Dutch grip on the region. Blocq would later write to his superiors to complain about 'the arrogance and viciousness' of the English, who had tried to interfere with people who were 'little more than our slaves'.

The Dutch defended their conduct by declaring that they had already contracted the producers from whom Saris tried to buy. This was what free trade required and allowed. In fact, the contracts had been imposed at gunpoint and had stipulated prices so low that none of the suppliers would have sold them any spices except under coercion. As the governor overseeing Dutch interests in the Moluccas wrote back to Amsterdam: 'We could not have prevented the English trade with the natives if we had tried to stop it on the strength of sworn contracts and agreements alone.' He blamed Islam for the failure of the suppliers to honour their contracts to sell their spices only to the Dutch, and advised that 'naked force' would have to be used henceforth to keep local rulers in line. The

Dutch ideal of the free sea was nothing but a legal fiction covering their brutal dominion over the Spice Islands. National trade craves monopoly.

Saris decided in early April to make for Tidore, one of the Spice Islands where the Spanish still held a base, to explore the possibility of forging an alliance to block the Dutch determination to monopolise the spice trade. The *Clove* approached the east side of Tidore, the deepwater side of the island that lacked easy anchorage. A strong swell pushed the *Clove* closer to shore than he intended, and the Spaniards responded by firing a shot to keep the English off. After a few more exchanges on both sides, the Spaniards fired a cannon without shot in it, a sign that they were willing to parlay. Saris answered in kind, and the Spanish commander, Don Fernando Besero, sent out two men in a boat bearing a white flag of truce to ask what nation they sailed under and what business they had. Despite finding themselves out on the water during a tropical downpour, the two Spaniards were unwilling to go aboard the *Clove*, preferring to get soaked rather than be seized and taken hostage. Saris had one Spaniard on board, Hernando, whom he had brought with him from Bantam in case there should be an opportunity to negotiate trade with other Spaniards. Hernando knew the two men in the boat, who were relieved to learn that this was not yet another Dutch ship come to harass them. They rowed back to shore to report to Don Fernando, who improved his welcome to the Englishmen by sending him his chief pilot, a man called Francisco Gomes.

Saris describes Gomes in his journal as 'a man of Good presence, with Compliment and, telling me I was welcom, offering me his assistance to bring me into the best anckoring place'. After the ship was moored and dinner shared, Gomes excused himself to go ashore and explain the situation to Don Fernando, who in any case could do nothing without contacting his superior on nearby Ternate, Jerónimo de Silva. Saris put Gomes ashore, 'for without his directyon they could doe nothing'. The pilot returned the following day with gifts of food from Don Fernando. Saris sent return gifts and an offer to trade food or munitions for cloves. Saris asked for a speedy response, for two Dutch ships were shadowing the *Clove* the whole time. Saris was banking on the Spanish being in desperate straits. They had wrested Ternate from the Portuguese seven years earlier, in 1606, but since 1607 they had been in a stand-off with

the Dutch over their prize. None other than Jacob van Heemskerck, cap-
tor of the *Santa Catarina*, had been to Ternate as early as 1601, though
another six years passed before the VOC set up a fortified base on the
side of the island away from the Spanish. Saris suspected that the running six-year conflict was bleeding the Spanish of supplies, especially
munitions. Their need could be his opening into the spice trade. If the
situation was as Saris supposed, Don Jerónimo did not let on. He sent
the Englishman a letter that, as Saris complained to Gomes, was 'nothing but compliments' and 'other idleness'. The governor invited Saris to
make a courtesy call at Ternate, but said nothing about trade.

Saris was offended by this politely worded suggestion that the English get lost. He had come to trade munitions for cloves, a straightforward
business deal, and thought the governor should have seen the wisdom
of going along with it. Gomes tried to keep the deal alive by explaining
that the main store of cloves on Tidore had been taken off four months
earlier, but that more had accumulated since. Approached properly, Don
Fernando might just be willing to make a deal without referring to Don
Jerónimo. Gomes would see what he could arrange and begged Saris to
delay his departure. Saris caught the drift: Gomes needed sweetening.
He and his senior staff put together a collection of navigational instruments that a pilot such as Gomes would be sure to appreciate. These did
buy Gomes's favour, but Saris began to suspect that some treachery was
being planned. He weighed anchor and crossed over to Ternate to call
on the governor in person, but Don Jerónimo made it clear that he would
not trade. All he wanted from the English, he claimed, were several pairs
of seamen's boots. Saris sent over three pairs gratis and switched to his
back-up plan: he set his course for Japan.

Saris's treatment by the Dutch was among the issues raised at the
second Anglo-Dutch conference in 1615. The first conference had been
under way while Saris was at sea. In fact, the day de Groot lectured
James I on the freedom of the sea, 13 April 1613, happened to be the day
Saris sent the boots to Don Jerónimo as a gesture of farewell to the Spice
Islands. Back in Europe, neither the Dutch nor the English knew what
was transpiring off Ternate, but they certainly did when the two sides
met for a second conference two years later at The Hague. De Groot took
the stage again, arguing that the Dutch position since the publication of

The Free Sea had been consistent. The English delegation cited Saris's journey as evidence of the double standard of the Dutch legal position, insisting that trade should be 'free as well for us as for yourselves'. The answer from the Dutch side came from de Groot. The contracts were valid, he declared, and the men who signed them and then tried to trade with Saris were 'perfidious' and not to be trusted. It was the English who were intimidating the local rulers, not the Dutch.

De Groot knew this was all a fiction. He was personally and legally offended at Blocq calling the Moluccans 'our slaves', but he was the VOC's lawyer and obliged to argue on his client's behalf. So he retreated to the high ground of contract law and insisted that 'he to whome another hath promised to deliver certayne commodities hath right to hinder the promiser from delivering them to any other'. In the narrowest of legal terms, the Dutch were simply enforcing contract. The issue had nothing to do with the freedom of the seas.

The second conference is regarded by fans of de Groot as a low point in their hero's career. He may have phrased the Dutch legal right to exclude the English by invoking the law of contract, but in reality the exclusion rested on the condition of virtual slavery of the people, and it was imposed and maintained solely by force. Van Ittersum concludes that de Groot could have had few doubts about the true nature of the VOC contracts. 'From the perspective of the indigenous peoples, the contracts were no longer voluntary agreements but cruel dictates that undermined their sovereignty and self-determination', she writes; 'the countervailing evidence at his disposal was never more copious and straightforward than in the case of Saris's voyage to the Moluccas.' This observation leads her to a broader conclusion. The VOC was an economic enterprise only in appearance. The Company existed to serve political and military considerations, tapping commercial wealth principally for those purposes. As she phrases it, de Groot's 'famous plea for freedom of trade and navigation seemed little more than a fig leaf for the VOC's naked self-interest'. But it worked. The Dutch retained their dominant position in the Spice Islands, and the English were left to squeeze into whatever modest, and ultimately unprofitable, niches they could find. Ultimately they would look to build their fortunes in South Asia and leave South-East Asia to the Dutch.

When John Saris arrived on the south coast of Japan on 9 June 1613, he was looking for Will Adams. Adams literally washed ashore in 1600, one of a handful of survivors who crossed the Pacific on a Dutch ship. The first Englishman to reach Japan, he spent the intervening thirteen years making himself useful to the newly ascendant shogun, Tokugawa Ieyasu, and gaining a thorough knowledge of the waters around Japan. The Company targeted him as an invaluable asset for inaugurating trade between Japan and Britain. They were right, and he was willing to serve their purpose. Being thoroughly embedded in Japanese society (he had a Japanese wife and children), Adams could navigate not just the coastal waters of Japan but its complex political and social waters as well, negotiating on the Company's behalf with everyone from the neighbours to the shogun. He also proved to be a skilled captain and pilot, later sailing cargo junks down the coast of China in the Company's service.[*]

Adams wasn't in the port of Hirado when Saris arrived, but a few other Europeans were. Through them he gained an introduction to the *daimyo*, or lord, of Hirado. Through this daimyo Saris was introduced to the senior Chinese merchant working in Japan, a man by the name of Li Dan. Li first comes into view in Saris's journal on 16 June 1613, six days after Saris's arrival at Hirado, as a prospective landlord for the factory Saris wanted to establish. He names him in his journal as Captain Andace, whom he identifies as 'Captain of the China quarter'. Captain Andace and Saris got on well, for he appears again in the journal a month later to bet Saris that the mysterious shipwrecked English pilot whom the English had yet to meet, Will Adams, would show up within four days. Saris lost the bet the following day, and was happy to, as he now had an Englishman who spoke the language and could help negotiate the intricacies of setting up a factory here for the East India Company.

Andace is how Saris records his name, but the English who stayed on in Hirado came to call him Andrea Dittis. The life and times of this

---

[*] 'Junk' is a technical term for a ship having a flat bottom rather than a deep keel, designed to transport cargo or people, usually both. The name comes from the Malay word *jonq*, doubling the use of an already existing nautical word in English, 'junk' being frayed rope no longer good for rigging.

man have been preserved for us thanks to the diary of the head of the Hirado factory, Richard Cocks. The portion of the diary that has survived opens on 1 June 1615. Cocks's entry for that day names not only Dittis but also his brother, Captain Whaw, to whom Cocks sent a gold bar as a christening gift for his youngest daughter. These are not obviously Chinese names, but the Japanese historian Seiichi Iwao has unlocked their riddles. Dittis is a Kyushu rendering of the Chinese name Li Dan (Di=Li, ttis=Dan), while Andrea was a 'Christian' name that Portuguese traders gave him. Whah (he also appears as Whowe) should properly be rendered Huayu: this is Li Dan's brother, Li Huayu.

The Li brothers hailed from Quanzhou, one of the two main coastal cities of Fujian province from which merchants and coolies sailed out into the trade networks of East Asia. As a young man Li Dan had gone to Manila, the base of Spanish trading operations in Asia, and amassed a fortune, or so Cocks gathered, but he had been forced to leave in the wake of the great Spanish massacre of Chinese there in 1603. Li fled to Japan, where he had lived for nine years before the English arrived. The brothers were older than Saris, who was about thirty-four when he sailed into Hirado, and probably just a little older than Cocks, who had already reached the then advanced age of forty-eight. (He was on this voyage late in life because his own business back in England had failed.) As Li Huayu died in 1619 and Li Dan followed him in 1625, both of natural causes, we might suppose they were on either side of fifty in 1613. They had put their nine years to good use in Japan. By the time the *Clove* sailed into Hirado, the Li brothers had installed themselves as leaders of the trading community of several hundred Chinese living at the south end of Japan. Both had residences in Nagasaki, where the Portuguese had opened trade and where the Dutch would end up once the Tokugawa regime closed the country, but Li Dan's main business was in Hirado, which also served as the base for Dutch and English commerce at that time.

The goal of every trader in the China seas was to get access to China, not something the Ming court was prepared to give. The official view from Beijing was that foreign trade could be conducted only as an adjunct of diplomacy. If foreign envoys came to offer tribute to the emperor, they would be permitted to conduct trade, and then only under official supervision. The court might relax the restrictions on foreign trade in times when it felt the borders were secure, but the maritime border rarely felt this way. Too many private operators worked the coast, especially Japanese, and when the court found itself unable to control piracy and smuggling, it preferred simply to shut down all private trade. This usually had the effect of only increasing piracy and smuggling.

What the Lis offered was access. They had business and official contacts inside Fujian province, or so they claimed, so bribes to the right people might just open trade opportunities there. China was the highest card in Li Dan's pack. As long as the English and Dutch in Hirado had no contacts inside China except through the Lis, there was money to be made by convincing them to finance a campaign of bribery. The expectation that Emperor Wanli would soon die – he did not oblige his weary subjects until the summer of 1620 – fuelled anticipation that his policies might be overturned, as they were so spectacularly half a century earlier when his father came to the throne as the Longqing emperor and rescinded the ban on maritime trade. If the Li brothers could get in early, they could manage the new wave of trade. Alternatively, if nothing changed, a turnover at court still raised the happy prospect that officials appointed under Wanli would be recalled, which would mean having to bribe the new slate of officials appointed by the emperor's son. This would then be a lovely opportunity for the Lis to go back to their foreign trade partners and levy a second round of major bribes. Under these conditions, the English factory in Hirado leaked money for years in pursuit of the chimera of the China trade.

Were the brothers telling the truth, about either their expenses or their prospects? Certainly it would have been hugely profitable to them to pry open trade with China and be the sole agents of the English in that trade. Li Dan did many times assure Cocks that he was dealing in good faith, promising that he would return all the money Cocks had given him if in the end the venture failed. In reality, however, the chances of accomplishing this were slim. Some counsellors in Beijing argued that opening maritime trade was the solution to both piracy and regional poverty, but court anxieties about domestic instability along the coast – plus the monopoly that the imperial household held over customs receipts – meant that the coast was rarely open for more than a year before it was closed again. The instability at court created a golden opportunity for entrepreneurs such as the Lis to take advantage of the benefits that pro-hibition brings to those who know how to flout it.

Cocks's eager but perhaps reckless stewardship of EIC operations pushed his relationship with Li Dan onto an ever more ambitious, and ever more expensive, footing. Gifts, loans and minor trade deals steadily grew into a partnership aimed at entering the China market. Cocks's commitment was cemented at the beginning of 1617 with a loan to Li Huayu of the substantial sum of 2,000 ounces of silver for a year at 20 per cent interest, plus a loan of half that amount on the same terms to another Chinese merchant in Nagasaki. More money would have to be thrown after that sum, for on 16 December, as the end of the loan period approached, Li Dan dropped in at Cocks's house with a gift and a letter from his brother saying nothing about the loan but asking for another 1,000 taels (ounces of silver) 'to be emploid about procuring trade into China'. The brothers would 'allwaies be answerable for it, whether it take effect or no'. Li also asked for the largest of the robes the shogun had given to Cocks to sweeten the Chinese officials, insisting again that for this too 'he would not forget to be answerable'. Cocks handed over his two best robes the next day, but had to wait until the lord of Hirado repaid some of his debt of 3,000 taels before he could send the silver Li Huayu asked for. Later that morning, Japanese officers on an English junk about to depart for Siam called on Cocks to hustle gratuities. He ended up having to fork out 65 taels on the pretence that the money would be used to buy wood in Siam. He knew that it wouldn't, but the junk would not set sail if he did not pay. At the end of that day's entry in his diary, Cocks wrote in exasperation: 'God blesse me out of the handes of these people.'

The trade to Siam was a side operation. The real goal was trade to China. In a letter to London dated 15 February 1618 Cocks professed optimism about his China venture. He made the point partly in response to having received two letters addressed to the emperor of China from James I, one friendly, the other threatening. The Company's Chinese translators in Bantam had been too nervous to translate them into Chi-nese, lest they be found guilty of the capital crime of offending the dig-nity of the Ming emperor. Cocks was happy to report that 'our China frendes', the Li brothers, were unfazed. They 'will not only translate them, but send them by such as will see them delivered'. Not both, actually; only the friendly letter. He gently reminded London that threatening letters would get them nowhere. The Li brothers assured him that 'there will nothing be donne with the king by force.'

Li Huayu's death two years later did not dampen Cocks's enthusiasm, for he was still confident that the surviving brother would 'prove the author of so happie a matter as to gett trade into China'. Li Dan, he wrote, was insisting that 'it is concluded upon, and that he expects a kinsman of his to come out of China with the Emperours passe, promesing to goe hym selfe with me in person, when we have any shipping com to goe in'. As late as 11 January 1621 Cocks put another 1,500 taels into the deal at a charge of 2 per cent a month, allegedly to match another 1,500 of Li Dan's own money, 'to procure free trade into China', with the assurance that, if the deal collapsed, the China Captain would repay the full amount with interest. Li did bring Cocks 2,000 taels in December 1622, saying he had intended to use it to pay down part of his debt, but that he had to give it to the lord of Hirado to keep him happy. Every promise seemed to compensate for the one before it and buoy up the prospects of the one to follow. The bet never paid out.

Readers over the centuries have loved the tales of Captain John Saris, Factor Richard Cocks and the stranded English pilot Will Adams. James Clavell thought Adams good enough material from which to write the novel *Shogun*. Historians have been less generous over what these adventurers actually did. They look at the labour and waste of the Hirado factory and regard it as a case of lost opportunity rather than of any real achievement. What everyone agrees on is that the China Captain took Cocks for a ride. I'm less sure. True, when the East India Company shut down its factory in Hirado in 1623, it was left holding debts amounting to 128,218 taels, of which 6,636 taels were debts that Li Dan had racked up. Cocks was strongly censured by the Company when he got to the regional office in Batavia, although Batavia left final judgement on his service to headquarters in London.

It was a judgement that never came. Cocks died at sea on 27 March 1624, en route back to England, in agony but of unstated causes. The EIC wrote off the losses and closed the account of his personal estate, which was valued at £300. This was a meagre showing for a merchant. Cocks had warned London as early as 1620 that the Japan outpost was unlikely to turn a profit and that he himself would gain nothing by it. As he put it rather bluntly, if 'other occasion amend it not, I shall, as I came a pore man out of England, retorne a beggar home, if your Wor[ships] have noe consideration thereof'. When the Court of Directors met on 24 November 1626 to consider a petition from Cocks's brother to inherit his estate, the directors felt that the Company owed nothing on Cocks's account: 'The Court related the debaust [debauched] carriage of his brother and the evill service performed by him at Japan, where he had lived long contrarie to the Companies mind and had expended 40,000 pounds, never returning anything to the Comp. but consuming whatsoever came to his hands in wastfull unnecessarie expences.'

The alleged loss is probably off by an order of ten. In the end, though, the Company settled Cocks's estate on his brother. The amount was small enough to mean nothing to them. Acceding to the petition, albeit after some chastising words, was better publicity for the Company. What catches the eye in the minutes is the accusation that Cocks was 'debauched'. Whatever vicarious pleasure Saris was getting from his collection of erotic books and prints, Cocks was managing to get in real time, or so the directors of the EIC thought. Cocks is extremely discreet about personal matters in his diary, although the occasional reference slips past his censorship. For example, he mentions attending a dinner at a merchant's house in Hirado on 8 September 1616 that ends with the host sending each guest one of the women who had sung and danced for them that evening. Little else of such exploits appears in the diary. We really have no idea what he got up to, and no way of confirming the Company's suspicions.

Read the diary often enough, though, and the name of a woman begins to recur. Cocks mentions Matinga for the first time on 2 August 1615, one month into the surviving portion of his diary. He notes on that occasion that he gave her 6 taels of silver to buy rice. On 25 September he gives her 2 taels that someone repaid him as a debt. On 29 December he gives her a piece of satin that cost 5 taels and a piece of taffeta that cost 1, so that she could make kimonos for herself and her two maids, Otto and

Fuco. Four days later, he notes that 'Matinga went into her new howse this day.' Towards the end of January he gives her 6 taels of silver 'to provide things against the new yeare' in her new house, as Spring Festival was approaching. These sorts of reference reappear every few months. It all seems rather anodyne until 22 April 1617, when Cocks notes that the lord of Hirado 'above a yeare past, sent me word he would geve me a howse rent free, which Matinga dwelled in, it being a matter of some 10 shilling or 2 *taels* per anno, but now goeth from his word and denieth it'. So Matinga had been living in the house that Cocks was given for his personal use for over a year; only now did he have to start paying rent. Other hints pop up in his diary, usually of a similar book-keeping nature. Thus, when he records paying the shoemaker, it is for two pairs of shoes for him – and a pair of wooden sandals for her. Not all references to her are actuarial. When he is away from Hirado, the only two people to whom he writes in Japanese while he is away are Li Dan and Matinga.

Cocks declines to name her as such in the diary, but glimmering through these casual references emerges the fact that Matinga was his Japanese wife. The relationship may have begun before 1615 and was still intact on 9 January 1619. That day he recites the gifts he had given to all his friends; the list registers a rich assortment of expensive gifts for her, including two painted screens. It also includes more modest gifts for her maid Otto. On 2 March it's all over. Otto wins her freedom by revealing that Matinga had been seeing half a dozen other men and can produce three witnesses to testify against her. The debauchery, if we want to call it that, was on her side, not his. This is the last we hear of Matinga. The marriage, such as it had been, was over.

The Company would not have grudged Cocks his Matinga. It was understood that the men sent abroad on Company business would establish relationships with women in the places they found themselves, regardless of whether they had wives at home in England. It was also understood that these relationships could be beneficial in all sorts of ways, connecting the newcomers to the society they entered and providing them with access to resources they might not otherwise enjoy. This was not a double standard, nor was it a double life; it was simply a double existence. It was an arrangement condoned, indeed encouraged, by all (except perhaps the wives left at home, though the trade-off was the promise of riches brought home). It was standard practice in Hirado, for Cocks in his diary repeatedly refers to the other Englishmen's women, usually in the context of having to give them gifts as well. He just declines to be explicit about his own.

London neither knew nor cared. What mattered to the Company was that the mission to Japan had failed. And it had been expensive. John Saris had picked up enough pepper and other commodities at ports throughout the region to make himself a reasonably wealthy man. All that came back to the Company at the close of the Japan venture was £1,000 in gold and £100 in silver, which was sent to the Tower to be coined. From a financial perspective Richard Cocks's legacy was nil.

Li Dan's financial legacy was worse. He outlived Cocks by a year and a half, continuing with the Dutch the same business ventures he had conducted with the English, which included holding out the promise that a deal with China was just about to be struck. His attempts to build a legitimate trading empire that linked the Dutch, the Ming and South-East Asia in a durable trading network would fail in 1624. The following summer he returned to Japan, defeated and in serious debt, and there he died. When word of his death reached Company representatives in Batavia the following January, they reported to London that he 'left a small estate to satisfie his cred[itor]rs and accordingly is distributed, out of w'ch you have your proportion in divers species' (that is, various precious metals). They promised a more concrete note to this effect, but none was ever sent. The EIC no longer had anyone on the scene in Hirado and could do nothing against Li's estate. The VOC was there, however, and saw matters with a clearer eye. The Dutch factor reported to his superiors that Li left nothing to cover his enormous debts, and that the English would not recover a penny of a debt he estimated as high as 70,000 taels. That was probably a tenfold exaggeration, designed to feed the Dutch penchant for making the English look as bad as possible. The fact was that the venture had failed financially. No more funds would be poured down that hole.*

---

* The Dutch found themselves in the same position in 1639, when the China Captain of Batavia (Jakarta) died with enormous debts the VOC could never recover. As the historian Leonard Blussé has argued, this was not a case of embezzlement. The

This is not, however, the end of the China Captain's story. Li did leave a legacy, not in the form of silver ingots but in the idea that there was profit to be made by controlling the trade around the South China Sea. So long as the Ming dynasty did not assert any sort of dominion beyond its immediate coastal waters, and so long as European power was not in a position to monopolise the trade, the ocean was wide open for whoever had the best ships, goods, weapons and knowledge of where and how to trade. The China Captain was not able to pull these strands of trade into a single system, but from among the ranks of the young men who flocked to Li's ships for work and adventure there arose one who did.

Zheng Zhilong was a handsome young Catholic convert who came over to Nagasaki from Macao on a ship owned by his uncle. Zheng caught the eye of the ageing Li Dan so powerfully that Zheng's earliest biographer suggests the youth became Li's lover before becoming his trusted protégé. Given the widespread practice of homosexual relations among the Fujian elite, it is a plausible explanation for why Li singled Zheng out from among the many young men working for him. Whatever the dimensions of their relationship, Zheng won Li's trust, worked closely with him in his negotiations with the Dutch and at the age of twenty-one took over the tatters of his boss's maritime empire when Li died.

By working with the Dutch and operating his own networks from Manila to Japan, Zheng Zhilong came to control much of the trade operating around the South China Sea. He stopped short of establishing his own political regime, especially when the Ming emperor offered to buy him out, effectively sidelining him as a competitor to power. In 1646, however, Zheng defected to the Manchus, who had conquered China two years earlier, and he lived out the rest of his life in Beijing as a well-rewarded hostage of the new Qing dynasty until he was executed in 1661 for the rebellion led by Zheng Chenggong, his son by a Nagasaki woman. The son had taken his father's place at sea and attained such supremacy on the water that he was strong enough to launch a campaign against the new Qing. Driving the Dutch from Taiwan in 1661, Zheng founded

Chinese merchant found himself caught short by a quagmire of deals and obligations that got deeper the more he tried to escape it by making impossible promises to his avaricious European clients.

a kingdom that he called Dong Ning ('Eastern Calm'), the first step to founding a separate dynasty. Although Zheng Chenggong died of malaria the following year, the Eastern Calm kingdom survived on Taiwan for another two decades before the Manchus crushed its power and annexed Taiwan to the Qing empire. This is how Taiwan became part of China.

The idea of an empire based more on water than on land was an innovation beyond Li Dan's imagining. He had learned that wealth was based not on what you owned but on how much you could trade it for, and he had turned that knowledge to personal advantage. The English and the Dutch would pay the price, yet so too would he, for his family lacked the organisation to benefit from their business and survive financial collapse. What Li did not imagine, and there is no reason he should have, is that the state could be the founder and protector of private commercial corporations. The Ming dynasty had never shown the slightest interest in serving as an advocate of trade. The Japanese lords in Hirado and Nagasaki, by contrast, were certainly interested in creaming the profits off his operations when they could and in using him effectively as an informal bank to fund their deficits, but that sort of racket could not produce a marriage between commerce and state in Japan either.

We shouldn't expect a beleaguered merchant cut adrift on the China seas to predict a future no one could see. The Dutch and the English states were putting some of their interests, and much of their futures, in the hands of private corporations. But no one, east or west, could anticipate the age of imperialism yet to begin. Li Dan had no reason to guess that, barely two decades after his death, his favourite's son would attempt to build a state regime by sea rather than by land. In Europe, though, the new deal between state and commerce stuck. Most people liked to think that de Groot was right about the sea being common to all, but even de Groot may have sensed that Selden knew better, that the coming game was not free trade but empire.

5

The Compass Rose

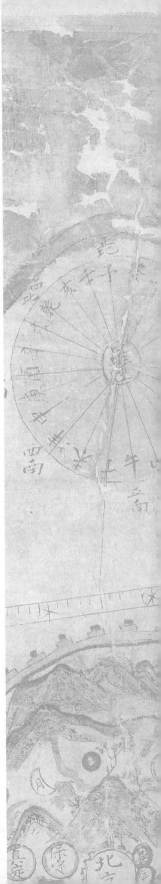

The strangest thing on the Selden map is the compass rose. It sits dead centre in the space above the Great Wall directly north of Beijing. Willow trees in Mongolia improbably droop their branches over it, and a flowering plum tree blossoms just as improbably to the west (Fig. 10). There is nothing strange about the actual compass. It is the standard Chinese version with twenty-four spokes, each with its directional name. It is a little odd that the Selden cartographer has inserted the word for 'compass', *luojing*, in the small circle in the middle, as though the image somehow required a label. (And it is indeed the cartographer's hand, not Michael Shen's.) No, what really makes the compass strange is that it shouldn't be there. Chinese maps never include a compass rose: none before the Selden map has one, and none after, until the European style took over in the twentieth century.

The compass rose is made stranger by its two companions, a ruler beneath it and an empty bordered rectangle drawn to one side. The ruler appears to represent one Chinese foot (*fen*), for it is graduated in units of ten inches (*cun*) and a hundred tenths (*fen*). It is not clear whether this foot is being used as a scale. A Chinese foot at the end of the Ming dynasty measured a little under 32 cm, whereas the ruler on the map measures 37.5 cm. In any case, Chinese maps never include a measurement scale. According to convention, neither the ruler nor the compass belongs here, yet here it is. And what are we to make of the empty rectangle? It looks like the sort of frame you might put around a title, but that isn't something Chinese cartographers did. When they gave titles to maps, they put them in a band across the top rather than intrude them into the space of the map. Occasionally they might insert a colophon to explain something – and in fact the Selden map has one such colophon on the left-hand side, which we will get to in due course – but it is always boxed by a single line, not a double.

If none of these three things naturally belongs on a Chinese map, then what are they doing here?

Let's begin not with this compass but with compasses in general. I think it's fair to say that everyone in this book owned at least one compass. John Saris owned several. We know this because, when he was trying to swing a trade deal with the Spanish at Tidore, he and his staff decided it would be politic to offer gifts to the Spanish pilot, Francisco Gomes. The master of the *Clove* contributed a half-hour glass and a half-minute glass to clock a ship's speed, a hemisphere to show the disposition of the lands and seas, and a 'dipsie' (deep-sea) line to sound depths. Richard Cocks threw in a quadrant 'for to observe the sun' and measure its angle to determine latitude. Saris gave him a quadrant as well, but he also gave him a sea compass. The English ship was clearly well supplied with such navigational equipment.

Neither Francisco Gomes nor John Saris could have known that the mariner's compass was a Chinese invention: Chinese sailors had been using them since at least the tenth century. Nor could they have known that the compass reached Europe through Persian pilots, who acquired this technology from Chinese in the thirteenth century, if not earlier, and thereafter passed it on to Mediterranean mariners. The one thing they might have heard is that the Chinese thought compasses pointed south rather than north. 'Compass' in Chinese is *zhinan*, 'pointing south'. South and north being on the same axis, it makes no difference which direction you choose to identify as orienting that axis. They certainly wouldn't have known that Chinese first utilised the tendency of magnetised matter to align with the earth's magnetic field by casting pieces of loadstone on divination boards to predict the future. Nor would they have known that the practice of casting pieces on a board anticipated the invention of chess, dominoes and, most surprisingly of all, playing cards. This complicated history was first put together by the great English historian of Chinese science, Joseph Needham, and published when I was eleven years old – long before I went to work for him as a research assistant two summers after going through Friendship Pass.

A compass was useful, but a compass alone did not enable a mariner to navigate unfamiliar waters. Local knowledge was essential, and for that you needed a pilot. Saris understood this necessity at several moments on his voyage. It would have been impossible to bring the *Clove* through the tricky entrance into the harbour at Hirado the first time without local pilots. In this instance he obtained the guidance of two masters of fishing vessels he picked up outside the harbour. It was an easy day's work for the Japanese fishermen, who received the handsome payment of 30 ounces of silver and a day's food for their trouble. These services were not always rendered voluntarily. As the *Clove* approached the treacherous waters around the Celebes on 31 January 1613, Saris overtook two transport vessels en route from Pattani, on the east side of the Malay Peninsula, to Ambon Island, to the east of the Celebes. Boarding the smaller vessel, he 'fecht the Master aboard to direct me through the straites'. He released him the next day to the larger vessel, rewarding him with a length of calico for his services and handing the master of that vessel 'a letter of favor to all English shipps he should meete withall'.

The dependence that European merchants had on Asian pilots could be a source of anxiety. In its first years in these waters the EIC mostly used Chinese pilots. This was partly in recognition of the fact that Chinese dominated the trade around the South China Sea. It was also a calculated political investment. Cocks reveals this in a letter to a Company colleague stationed in Pattani, advising him to 'use all Chinas kindly & w'th respeckt'. His advice was based on a single overriding premise: the Company wanted access to China. The Company's entire strategy in Asia hinged on it. Cocks did not want any Englishman to give any Chinese an excuse to reject the Company's request for direct trade. As he whispered to his compatriot in Pattani: 'I am certenly enformed that the Emperour of China hath sent spies into all these partes of the world where the English, Dutch, Spaniardes & Portingales doe trade to see their demeanors & how they behave themselves towardes the Chinas nation.' The notion of a master espionage plan radiating from the court may be a little far-fetched, but the endless difficulties Europeans faced trying to do business with China, and not getting anywhere close to that ambition, fed such suspicions. Cocks was desperate for a break-through on the China trade and didn't want anything to spoil it. He warned the man in Pattani not to leak this story, lest the Chinese get wind that the English knew what they were up to. 'This I wrote you is no fable but truth,' he declares at the end of his letter, 'yet keepe it to yourself.'

Pilots of the Ming dynasty have remained steadfastly anonymous. I have not found a Chinese Francisco Gomes lurking in sources of the early seventeenth century. What keeps them from our acquaintance is the secrecy with which Chinese merchants surrounded their operations. Ship owners would have recorded pilots' wages in account books, but businesses were extremely careful not to allow their books to escape their control. A competitor could not be allowed to peer into the state of a business's finances – even worse, a tax official. With only one exception to my knowledge, there are no account books of commercial firms that survive from before the eighteenth century, and of merchant trading firms, none at all. This is why we cannot name a single Ming pilot. They did not write about themselves, and they remained off the radar screen of those who kept public records.

But there is one man who so loved the sea that he set himself the task of gathering every scrap of information about ships, routes, compasses, piloting and the tricks and technologies of sailing, all of which he compiled into a book he finished in 1617 and may have published the following year. Zhang Xie was neither a pilot nor a ship's captain. His early path did not point him towards the sea. Like any other son of a family that could afford to sacrifice his labour, he had to study for the examinations that led to the civil service and a career in the imperial bureaucracy. His course through the thickets of Confucian texts and examinations was set for him. It started in the local county school in his home town of Zhangzhou, the most southerly prefectural capital in Fujian province; it went north to the provincial capital, Fuzhou; and beyond that it stretched as a bright but crowded road that ended in Beijing. Zhang got to Fuzhou and earned the degree of *juren*, Recommendee, in 1594. But the path to Beijing proved too narrow and difficult for him to rise to the top degree of *jinshi*, Presented Scholar. The provincial degree became the ceiling through which he could not break. It was sufficient to qualify for a post lower down in the civil service, perhaps as a county vice-magistrate, but inadequate to compete with the Presented Scholars.

One ancestor four generations up another branch of the family had risen through the exam system and the civil service to become a vice-minister, but no descendant of his was able to match that record. Zhang Xie's father, Zhang Tingbang, had earned the title of Recommendee in 1572. That had been sufficient for him to be appointed as a county magistrate. He did well enough to earn a promotion to vice-prefect, but his unwillingness to fawn over the prefect got him fired in his early thirties. That was the end of his career in the civil service. Zhang Tingbang had to go home to Zhangzhou. All we know of his subsequent life in enforced retirement appears in a brief biographical notice in a Fujian miscellany. The biography records that he had a residence on the west side of the stateliest Buddhist temple in the city, which he called the Hall of Stylish Elegance, and that he converted it into a clubhouse for a poetry society in 1601. It also tells us that he liked boats, for he ended up living on a houseboat, spending the rest of his life afloat on the rivers of his native place.

We shouldn't read too much into given names, but names do have meanings. Tingbang means 'court proclamation', Xie means 'harmony'. If ever two names announced a family's change in direction, these of father and son did. Zhang Tingbang's family clearly wanted him to get to Beijing, but Tingbang didn't impose that ambition on his son. Although he passed the provincial-level exams at the reasonably young age of twenty, Zhang Xie had no desire to repeat his father's experience and make the compromises required for getting ahead in a bureaucratic career. He went travelling instead, visiting all the great sites throughout the country and hobnobbing with some of the leading lights of the age. Everyone who mattered got to know Zhang Xie and count him as a friend. Once back in Zhangzhou, he took the moniker Coastal Scribbler and turned to writing. He published fifteen books in all, mostly poetry, all of it lost – with one startling exception, without which he would have been entirely forgotten.

Dong xi yang kao ('Study of the Eastern and Western Seas') is the only account we have of Chinese maritime endeavours in the South China Sea. Despite its insipid, academic title, it is one of the remarkable books of the age. It started out as an assignment. The county magistrate of Haicheng, Zhangzhou's port on the coast, wanted some kind of record about the maritime world that started at the lighthouse on Gui Island, beyond the far end of Moon Harbour, and stretched outward to the world. A *gui* was an ancient jade object signifying imperial power, cosmically shaped round at the top and square on the bottom, as were heaven and earth. The magistrate was about as far from being a seafaring man as any state official could be, but he needed to know something about wherever it was that half of the county population went and what they did when they got there. Zhang Xie could have completed this assignment in a perfunctory way. Instead, he turned it into a project that absorbed his immense energy, sending him down to the docks and into the archives to assemble the knowledge he needed to compose a complete picture of the maritime world. The manuscript might have languished unfinished and unpublished but for an enthusiastic prefectural official in Zhangzhou who found out about it. As this official noted in his preface to the book, Haicheng was a *shuiguo* – a water kingdom – where 'daily necessities come from across the sea, luxuries are foreign products, and kids in the villages who can barely talk can translate foreign languages!' No county in China was quite like this, and it needed a record that could attest to just how intimately the county was bound up with the maritime world.

I think of Zhang as energetic, although it is not easy to catch a glimpse of the man behind the book. He writes in a straightforward style but occasionally turns an elegant phrase as well, suggesting that he enjoyed the craft of writing. He states what he likes and does not hide what annoys him, chews out incompetent record keepers and lax officials, and is merrily contemptuous of those who have tried to write about the sea and got everything wrong for want of direct experience. Zhang doesn't reveal whether he himself ever embarked on anything more nautically challenging than his father's houseboat, but he does say that he spent a lot of time talking to sailors in Haicheng to learn as close to first-hand as he could about seafaring. He is also refreshingly open-minded about foreigners. When someone demanded to know why he was glorifying such unpleasant folk as Japanese and Dutchmen by writing about them, he responded that this wasn't at all what he was doing. He was writing about those who obstructed commercial vessels and who happened to be Japanese or Dutch. Circulation, not ethnicity, was the important thing. He would have agreed with de Groot about the importance of the free sea, if he had known about him.

Zhang Xie reserves his highest praise for the crew member known as the Fire Chief: the man we call the pilot. Fire Chief seems an odd title. Joseph Needham guessed that it had to do with the practice of permitting only the lead ship of a convoy, where the pilot would be, to carry fire. I suspect not. Chinese cargo junks did not regularly sail in convoy; moreover, a ship's master would be foolish to go to sea without a pilot even if he was sailing in convoy, since his vessel might well get separated from the others.

A better guess came to me when reading a passage in the memoirs of William Dampier. The English freebooter (the gentleman's word for a pirate), who will make a second appearance in the epilogue of this book, got a chance to board a Chinese junk and described what he saw in strikingly positive terms. 'She was built with a square flat Head as well as Stern,' Dampier writes, 'only the Head or fore Part was not so broad as the Stern. On her deck she had little thacht Houses like Hovels, covered with Palmeto Leaves, and raised about 3 Foot high, for the Seamen to creep into.' Below decks, Dampier was even more impressed with the construction and organisation of always scarce shipboard space. 'The Hold was divided in many small Partitions, all of them made so tight, that if a Leak should Spring up in any one of them, it could go no farther, and so could do but little Damage, but only to the Goods in the bottom of that Room where the Leak springs up.' This was also where the merchants accompanying their goods found shelter. 'Each of these Rooms belong to one or two Merchants, or more; and every Man freights his Goods in his own Room; and probably Lodges there, if he be on Board himself.' Up on deck, Dampier praises the mast and rigging. The main mast 'seemed to me as big as any third-rate Man of Wars Mast in England', third being a high rating in the British system. It was 'not pieced as ours, but made of one grown Tree', he writes, adding that 'in all my Travels I never saw single Tree-masts so big in the Body, and so long, and yet so well tapered, as I have seen in the Chinese Jonks.'

At the end of his visit Dampier stuck his nose in the cabin at the stern of the ship, 'wherein was an Altar and a Lamp burning, I did but just look in, and saw not the Idol'. Dampier knew to expect a shrine on board, for all Chinese ships carried an altar to Mazu, the Empress of Heaven, to whom sailors prayed for protection at sea. What he did not know was that this was where the pilot kept his compasses. And thus we have an explanation for why the pilot was called the Fire Chief. It was his duty to keep the vessel on course, but it was also his duty to keep the lamp on the altar lit, lest the attention of Mazu and any of the other gods receiving offerings wander from the vessel and leave the sailors defenceless against the malevolent forces that were always on the verge of imperilling sea travellers.

Zhang Xie was just as impressed with the vessels that the sailors of Moon Harbour launched onto the waves. 'The larger ships are over ten metres in the beam and over thirty metres long. Even the smaller ships are six metres in the beam and twenty metres long.' These vessels represented a huge investment. 'The cost of building a ship can run to over a thousand ounces of silver. Then when they return every year, they have to be completely overhauled and refitted, which cannot run to less than five or six hundred ounces.' What made them commercially viable was the skill of the pilot. Zhang testifies that his decisions overrode the captain's: 'Though the deeps across which he traces his watery course are vast, he is listened to in all things affecting the command of the ship.' His authority as navigator was entirely knowledge-based. 'He knows there are regularities in cloud formations and the movement of winds, and with this knowledge he ploughs through the waves for ten thousand *li* and is never once fooled into taking the wrong course.' And the key technology in all this was the compass. 'Relying on it,' Zhang writes, mariners

> may grope their way forward in the gloom and yet have complete knowledge of whatever part of the ocean they are in and what dangers they have to look out for. Indifferent to the storms buffeting them, they remain at ease in their places. Despite sharp winds and crashing waves, they voyage on as though everything were normal. With their long experience at doing this, they sail as though they are walking on level ground. At a glance they can figure out whatever it is they need to know.

It wasn't quite that easy. Deducing a ship's position requires knowing its direction and its speed. Direction was easy: you could read it off the compass. A Chinese pilot was expected to check his bearing every 20 nautical miles, taking readings at both the bow and the stern to ensure he was not getting a distorted reading. Speed was more difficult. It could be calculated by dividing distance by time, but neither distance nor time could be exactly measured at sea. Time could be approximated by tracking the passage of the sun across the sky or by burning a fixed amount of incense for which the burn rate was known. Distance was the bigger challenge once a ship sailed out of sight of land. For purposes of general calculation, Ming navigators estimated that a ship could sail four-fifths of a 'stage' – a land distance of roughly 30 kilometres – in 1¼ *geng* or 'watches', of which there were ten per twenty-four-hour day. At that speed a ship would cover 24 kilometres, or roughly 15 nautical miles, in 2 hours and 24 minutes, which calculates as a speed of 6¼ knots (nautical miles per hour).* A pilot could adjust his estimate of speed by dropping

---

* My estimate of 6¼ knots is slower than the range scholars usually give, from 12 to 20 knots. My limited experience of sailing suggests to me that their speeds are unrealistic. A wide, flat-bottomed junk would do well to sail at 6 knots. In the introduction to his edition of the Laud rutter, the historian Xiang Da found that, when actual distances are worked out against time travelled, a watch could be as short as 31 *li*, roughly 10 nautical miles, indicating a speed of about 4 knots.

a piece of wood – what European mariners called a 'speed log' – from the bow and pacing with it to the stern, then multiplying the time taken by the ratio between a given distance (one 'stage') and the length of the ship. The result of this calculation was not terrifically exact, as ships were subject to tide, current and swell, any of which could confound measuring how far it actually travelled over the ground.

All of this is to say that a compass on its own was not enough to navigate blue water where the distances were too vast and the markers too infrequent. To fill the knowledge gap, pilots turned to written records that explained how particular routes went. These records could be anything from point-form notes to rough sketches to full-blown sea charts. John Saris called them 'platts' or 'plotts', an archaic usage that survives today in the expression 'plotting a course'. Saris refers to them several times in his journal, usually to complain of their inaccuracy. More intriguing is an account in a letter Saris later received from his kinsman Edmund Sayers describing a difficult journey from Siam to Hirado. When the junk's incompetent Chinese pilot became so ill that he couldn't 'creepe out of his cabbin', Sayers had to take over with no idea of the ship's location. He then discovered that one of the cooks on board had what he called an old 'platt'. He was able to read it well enough to set a bearing that brought the ship to Japan six days later, with barely five men standing.

Almost no platts survive. What do survive are route guides: the English called them 'rutters'; the Portuguese, *portolanos*, literally 'lists of ports'; and the Chinese, *zhenjing*, 'compass manuals'. These hand-copied documents furnished pilots with the information they needed to get from point A to point B – or point *Jia* to point *Yi*, as Chinese would say. Each route starts from a specific port, proceeds as a sequence of compass bearings and numbers of watches that a ship should hold that bearing, and ends at the port of destination. Compass manuals were carefully guarded craft knowledge that was passed down within families and never leaked to outsiders. Accordingly, none survives in its original form. But a few came into the possession of writers in the Ming dynasty, and they copied them into other books. Zhang Xie's *Study of the Eastern and Western Seas* is one such.

Zhang derived the data in the long central section of his book, which data he calls 'compass knowledge', from several rutters he was able to get his hands on. And if it was hard to acquire rutters, it was even more difficult to construe their contents, for they were written in a telegraphic way that only experienced navigators could understand. 'The old maritime compass manuals of the navigators are all written in colloquial jargon and not easy to interpret', Zhang complains, 'so I sort of translated them into proper language. The parts that were reliable and worth recording I kept and dressed up a bit.' He also decided to reorganise the routes into a more systematic system. The data 'being all confused and disconnected the one from the next, I fused them into a unity. I have organised everything by route. Where a route forks and enters a port, I have noted it as a subsidiary route going to such-and-such a country, after which I continue on with the main route, doing the same thing at the next fork.' Zhang was pretty satisfied with the result, declaring, 'I managed to fit the entire ocean onto a scroll no longer than a foot' – a metaphor for his book (there is no map) – 'so that you can more or less see it all at a glance'. The mess of reality under his hand was transformed into a coherent, systematic account of all the routes around the South and East China Seas. It would have been nice to have the original documents in all their messy reality, but we don't. But better Zhang's version than nothing at all.

One manuscript of compass routes does survive. It is called the Laud rutter. If Zhang's *Study of the Eastern and Western Seas* is a step away from an authentic pilot's manual, the Laud rutter is half a step closer. William Laud donated this manuscript to the Bodleian Library in 1639 (Fig. 11). That the Laud rutter and the Selden map, each unique, should both end up in the same library seems too incredible to be true. But then it is difficult to imagine where else a compass manual and a chart of the China seas could possibly have survived except in a place removed from the trade and politics they document. It is possible that they reached England together, but they arrived in Oxford separately, and that is as far back as we can take them. What we can say is that they were brought by the same forces: the spice trade, global maritime connections and a keen interest in Oriental knowledge among scholars of the day.

———

Archbishop Laud was a controversial figure to his contemporaries and a complicated person to himself. A celibate vicar of modest upbringing,

Laud rose to the powerful post of Bishop of London under James I and was appointed Archbishop of Canterbury under Charles I. The further he rose in the service of the increasingly unpopular Stuart kings, the more staunchly he supported them, even to the point of upholding James's claim to divine right. Laud was a shrewd man, instinctual in his ability to manipulate others, but he was also vain about his own judgements and thus easily gulled by his own certainties. He had risen from very modest circumstances, which may explain in part his political ambition but not the actual political course he ended up taking. Someone more devoted to his own personal advantage would have figured out how to pull out of the downward political tailspin that led in 1641 to his incarceration in the Tower. He was alert enough to see what was coming two years earlier, when he donated the bulk of his Oriental manuscripts to the Bodleian rather than let them fall into the hands of his enemies. But he was not so ambitious as to abandon the principles by which he powered that ambition, which included the sanctity of the king and the preservation of the Church he headed. Three and a half years later he would be beheaded by the Puritan extremists he disdained.

Laud's disconcerting combination of theological subtlety and political appetite made him revered by some, loathed by many and avoided by most when they didn't owe him anything. John Selden was among the latter. Not being a churchman or interested in religious matters, he had nothing to do with Laud until his second imprisonment, when the archbishop chose to position himself as Selden's patron. It was rumoured that Laud worked out the deal with Charles to release Selden in exchange for the publication of *The Closed Sea*, although that rumour may have originated with Laud himself. Some kind of deal was struck, to judge from Laud's diary entry for 2 February 1636: 'My nearer care of J.S. was professed, and his promise to be guided by me; and absolutely settled on Friday after.'

Selden may have misled himself into thinking he could exploit Laud to find a middle way between the king and bishops clamouring for their prerogatives on the one side and the republicans in Parliament who wanted to tear down those privileges and erect their own on the other. A token of how close they may have become can be seen today in the History of Science Museum in Oxford. John Selden somehow came to own two Persian astrolabes for measuring the angles of the sun and stars to determine location, both of them beautifully worked in metal. Their exact provenance is unknown, although they were probably acquired in North Africa. One of these he presented to Laud as a gift. Both bequeathed them to the Bodleian Library, albeit separately. The Bodleian later lent them to the Ashmolean Museum as a more suitable place to house cultural artefacts. The Ashmolean has moved, but the astrolabes are still on display in the original Ashmolean building, which is now Oxford's History of Science Museum.*

The friendship between Selden and Laud, if it was ever that, did not hold. Laud was too extreme in his vigilance on behalf of the Church and of himself as its chief representative, and too unwilling to ease away from controversial positions that a more politically flexible archbishop might have gently abandoned. Selden admired him for his work as Chancellor of Oxford, and especially for his support for Oriental studies. They agreed about the intellectual poverty and moral vacuity of Puritanism. Laud would have found amusing Selden's snide comment that 'the Puritan would be judged by the word of God: if he would speak clearly, he means himself.' But Laud could not have accepted Selden's rejection of appeals to divinity in purely human affairs. And he surely must have been uncomfortable with Selden's puncturing of legal arguments for the divine right of bishops in his *Historie of Tithes*.

By the 1630s, Selden also faulted Laud on more practical issues. He felt that Laud went out of his way to support unpopular policies, notably the tax known as ship money. Charles I imposed this tax without the consent of Parliament, which would never have given it, in order to pay for an increased naval presence in the North Sea and the English Channel. His critics suspected Charles of wanting a navy he could use to intervene in the then ongoing succession struggles in Europe known as the Thirty Years War. If Laud assumed that Selden's arguments in *The Closed Sea* would strengthen Charles's right to levy taxes in defence of his dominion over the seas around Britain, he was mistaken. Selden held that the

---

* Here is a piece of pure speculation: as Selden had two Persian astrolabes and presented one to Laud , might he have owned a Chinese map and a Chinese rutter together, and similarly give one of the pair to Laud?

power to tax lay entirely with the people; kings and bishops enjoyed that power only to the extent that the people gave it to them.

By the time Laud wrote his last letter to Selden in November 1640, assuring him that he would persuade Charles to drop the ship money tax, the issue was beyond his power to reverse. Selden was appointed to the Parliamentary committee in 1641 formed to draw up charges against the archbishop, although he may have been able to dodge what would have been for him an unpleasant duty. When Laud was finally put on trial three years later, Selden helped him by furnishing historical documents that he requested for his defence. But he had no power to stop the train of events from leading where it did: to Laud's execution the following January. Selden did not agree with Laud's views, but he agreed less that executing him could ever be regarded as a legal act of justice.

Laud fervently believed in the importance of learning and of preserving knowledge for the future, even if he was not himself a scholar. And so he was an active chancellor, raising funds, financing buildings, funding Oxford's first lectureship in Arabic and arranging donations to the Bodleian Library. He was a major donor himself, giving the library over a thousand manuscripts. A quarter of these were in Oriental languages, including Chinese. As Laud wrote in 1634, such materials contained 'a great deale of Learning and that very fitt and necessary to be knowne'; without them, he complained, 'very few [students spend] any of theyr time to attaine to skill either in [Arabic] or other Easterne Languages'. He was determined to mend this lapse. And so a Chinese manuscript, the contents of which were entirely inscrutable to him, found its way into the Bodleian Library.

———————

The Laud rutter is not a working pilot's manual. It is a copy of one, edited into its present form by an anonymous editor who acquired the original but who, like Zhang Xie, decided to clean it up. In his preface the Laud editor makes the conventional claim that the method of direction-finding using a compass was invented by the Duke of Zhou, a much-mythologised regent in the early years of the Zhou dynasty (eleventh century BC) who is credited with everything good and noble in Chinese society. The editor then explains that rutters tend to be a jumble of information, some of it written, some of it drawn, much of it inconsistent, so he set himself the task of imposing order on it.

What soon becomes apparent as the reader proceeds through the handbook is that the Laud rutter is based on a record of the routes sailed by the imperial eunuch Zheng He in the fifteenth century. Zheng was charged by the third Ming emperor to go out into the world and inform all states with which China had relations that a new emperor had ascended the throne (which he did illegally by usurping power from his nephew, although that was not to be mentioned) and that he expected them to acknowledge his enthronement by sending tribute. This diplomatic remit resulted in half a dozen voyages between 1405 and 1422 first to South-East Asia and then into the Indian Ocean as far as the east coast of Africa. My first reaction to this discovery was disappointment. I wanted evidence of maritime trade, not of diplomatic missions. In fact, once I compared the routes in the Laud rutter to those on the Selden map, I found that the two enterprises ran on the same tracks. This coincidence was not because the eunuch fleets opened these routes for others to follow; it was because everyone used the same sea lanes before, during and after the Ming dynasty. Zheng He simply sailed his treasure ships, as they were called, where merchant seamen sailed their cargo junks. Most of what is in the Laud rutter is thus standard navigational lore, unaffected by the fact that imperial fleets sailed the same routes.

The rutter itself begins with a long prayer that a pilot was expected to make before consulting the compass. Addressed to the Yellow Emperor, the Duke of Zhou, six historical sages and every master who had ever successfully piloted a vessel across the ocean, the prayer asks for a safe and profitable voyage. There follows an explanation of how to use a compass. This would not have been a dry-pivot compass of the sort we are used to, in which the needle is attached to a post. Instead, it was a floating compass, which consisted of a shallow basin whose rim was marked with the points of the compass. The basin had first to be filled with water, then the needle placed on its surface so that it floated. John Selden believed he had acquired just such a compass, as he notes in his will: 'a Sea Compasse of their making and Devisione' (division). If this is the compass that is now on display in the History of Science Museum along with the two Persian astrolabes Selden and Laud donated separately (Fig. 12) then we know that Selden was mistaken. His was not a sea compass a pilot would take to sea, but a dry-point compass of the sort a geomancer used to site a building or a grave. Sea compasses floated on water.

Pouring water into the compass is tricky, the rutter explains. 'To fill it with the water on which the needle is set down, it is essential to use *yang* water and not to use *yin* water. What are *yin* and *yang* water? *Yang* water is water that trembles when the wind brushes it. *Yin* water is water that subsides when the wind falls.' I am not confident of what that translation actually means. Could this be telling the pilot to use salt water rather than fresh, so that it would help the needle to float? The next step is: 'Lay the compass basin level and orient it towards the south. Then it is necessary to set down the needle pointing to the mansion of Heaven.' Heaven here is not all of heaven, but one of the twenty-eight lunar mansions that Chinese astrology identifies as forming a circle above the horizon through which the moon passes, Heaven being situated at 315° from north. As the text explains, 'The mansion of Heaven is the chief of the twenty-four directions. Heaven governs the disposition of calamity, which is why it is necessary to start the needle from there. The needle should then indicate the direction, but don't let it sink.' The point of this advice seems to be to avoid laying the needle in the direction you think you are going. Instead you should start off at an angle that is always random to the ship's bearing but consistent to the north–south axis. This procedure ensures that the needle is not preset but 'finds' that axis on its own.

Precision in compass use was essential. 'If you make an error of a tiny fraction,' the Laud editor declares a little pompously, 'you may miss your destination by a thousand miles, by which time it will be too late for regret.' He provides a few other snippets of information on the use of the compass. He lists the points of the compass at which the sun and moon rise and set each month. He also includes four methods of calculating relationships among compass points. This was probably old lore that pilots used to train apprentices in the mental agility they would need when calculating points on the compass. Of these, the only method navigators learn today is what the rutter calls the Four Directions Method. It consists of a simple set of three four-character lines that reminds a pilot how to calculate reciprocals. These are the compass points on the opposite side of the circle (180° is reciprocal to 360°,

150° to 330°, 120° to 300° and so on). Part of the challenge of learning the Chinese compass is that points on the circle are not numbered but named. For a Chinese pilot, 180° is called *wu*, and its reciprocal is *zi*, which we know as 360°. (These two positions have counterparts in the organisation of times of day, '*zi* night' being midnight and 'mid-*wu*' being noon.) The reciprocals at right-angles to these are *mao* and *you*, which for us are 90° and 270°. All we have to do is add or subtract 180 to calculate the reciprocal, whereas the Chinese pilot had to memorise their names and relationships so that they can pop into his head without having to think them through.

I haven't found a Ming source that explains how the naming system works, but the eleventh-century author Shen Gua gives a good description in his monumental compendium of notes and observations entitled *Mengxi bitan* or 'Pen Conversations from Dream Brook'. It comes up in the course of his describing a mapping technique he calls As the Bird Flies. We attribute flying in a straight line to crows; Shen thought it was true of all birds. As the Bird Flies was developed to compensate for the problems of mapping an irregular surface that cannot be paced out. We don't need to go through the entire method, which is in any case rather poorly explained. But in the course of explaining it, Shen describes how to use a compass to draw a map.

Select any point on the landscape, Shen explains, and draw straight lines from that central point in the four cardinal directions (N, W, S, E). Next, double these by adding the four ordinal directions (NE, SE, SW, NW) for a total of eight. Up to this point, the Chinese and Persian/European compasses are the same. Both traditions refer to the four cardinal and four ordinal directions – and both call them the eight winds. (To judge from the journal Will Adams kept when he sailed from Japan to Tonkin in 1617, the Chinese names for the eight winds were among the first words a European mariner wanted to learn.) Beyond those eight winds, the systems diverge. The Western compass bisects: you divide the angles of the eight winds into sixteen points (half-winds) and then subdivide them into thirty-two (quarter-winds). Each of these thirty-two winds has a distinctive name in Italian, and every student of navigation had to be able to recite all of them in sequence, an exercise known as 'boxing the compass'.

The Chinese boxed a different compass. To make sense of their system, we have to go back to the eight cardinal and ordinal directions and start again. These eight winds divide the circle into 45° wedges. The logic of bisection for the European compass produces half-winds every 22½° around the circle, and quarter-winds at every 11¼°. The Chinese system trisects rather than bisects. It does this by adding two lines on either side of each of the eight lines, producing a central point from which radiate twenty-four lines rather than sixteen, each pointing in a direction that has a unique name. Instead of producing four wedges of 11¼°, trisection creates three wedges measuring 15°. The Chinese compass then trisects again, producing points at every 5°, unlike the Western compass, which bisects 11¼° into wedges of 5⅝°. The European system is geometrically clean but arithmetically messy, the Chinese system elegant and simple. To go back to the metaphors of direct flight, why trouble yourself with figuring out how crows fly at 5⅝° when you could have birds do it at 5°?

The naming system for the basic set of twenty-four compass points is not simple, even for Chinese. This is because it combines names taken from three different Chinese counting systems: base 8 (the eight Trigrams of the *I Ching* or *Book of Changes*), base 10 (a sequence known as the Heavenly Stems) and base 12 (a complementary sequence known as the Earthly Branches). The compass does not use all the names in all three counting systems, which would be too many for the twenty-four points that needed to be named. Instead, it takes four of the trigrams and eight of the stems but all twelve branches. That mix means mastering a lexicon of twenty-four names in fixed sequence without an internal logic that I've been able to discover. After mastering the twenty-four names, things get easier. The next trisection produces a total of seventy-two compass points. Rather than give the additional forty-eight points forty-eight new names, the Chinese system gives them names based on which two of the twenty-four they are between.

This part of the system is completely regular and easily understood. Let me demonstrate how this works by giving an example. If you are sailing due north, you are travelling in the *zi* direction at 0°. Turn 15° east and you are on a *gui* course. Vary your *zi* course by 5° to the east and you are said to be sailing *guizi*, meaning to the *gui* side of *zi*. English nautical convention would say that you are going '*zi* by *gui*'. (Think of Alfred Hitchcock's *North by Northwest* – although, as has often been pointed out, there is no such direction; the correct bearing is 'north by west'.) Shift another 5° to the east, and you will be travelling *zigui*, or '*gui* by *zi*'. That is, you will be on the *zi* side of *gui*, or 5° west from *gui* at 15° – which is to say, 10°. Turn another 5° to 15°, and you are on a straight *gui* direction. Keep moving around the compass by 5° increments and the pattern repeats itself with the next pair of names. If none of this makes obvious sense, it doesn't require mastering. Readers keenly interested in how the system works will find it laid out in the first appendix.[*]

———————

I observed at the start of this chapter that the most important thing to know about the compass rose on the Selden map is that it shouldn't be there: roses did not become part of Chinese cartographic practice until the twentieth century. Hyde registers no surprise at seeing it there, but then why should he, as he had few other Chinese maps against which to compare this Chinese map? He would have known them from European maps, of course, and yet as it happens, compass roses were disappearing from European maps. By the time Hyde and Shen were annotating the

———————

[*] Thomas Hyde was interested in the compass. Rather than annotating it on the map, he had Michael redraw it on a separate sheet of paper and mark all twenty-four points. Beneath it, he has written a note explaining that the naming system is also used to designate hours of the day and names of the months in a recurring cycle of sixty. On the back of the sheet Michael has written out the entire cycle, both the characters and their romanisations, with the eight trigrams in the middle. Hyde later added a note saying that the entire cycle 'may be seen at large in *Dr Hyde's History of the Religion of the Old Persians* printed at Oxford 1700, in 4to'. As indeed it can: the cycle of sixty names that he learned from Michael appears on a leaf of paper inserted between pages 218 and 219 of his magnum opus, complete with Chinese characters. They don't really have a place in that book; yet Hyde was so determined to publish this material somewhere that he shoe-horned it in. With this note we once again run up against Hyde's dream of publishing the first English book on the Chinese language. He even had his printer mock some of this up on one printed sheet, possibly in the hope of attracting a commercial publisher for that book on Chinese that he never got into print.

Selden map, the compass rose had become a hold-over from a particular phase of European mapping known as the portolan chart.

Portolan charts were charts of the places documented in the old portolanos or rutters. The earliest surviving example dates from the thirteenth century. They continued to be used until the seventeenth, when mathematical cartography pushed aside the method on which they were drawn. Rather than mapping a route on the water, portolan charts mapped the coastlines along which sea routes ran. The challenge a portolan chart-maker faced was how to move from section to section and maintain a consistency of scale and orientation. Compass roses were the answer. These were arranged in a large circle, ideally sixteen of them, to serve as nodes of reference for every point on the map. If all points could be calibrated by their magnetic angle to every rose, then the map that resulted would be perfectly correct. The lines radiating from the compass roses are known as rhumb lines: that is, lines that maintain a constant magnetic direction. In the old portolan charts these lines crisscross the surface of the map in a dense, Lilliputian web. Conventionally the lines at the eight winds were drawn in brown or black ink, at the eight half-winds in green, and at the sixteen quarter-winds in red. These rhumb lines are all anchored at some point to a compass rose, whether it is actually shown on the chart or implied somewhere off it. There could be up to sixteen of these nodes (like the sixteen half-winds), and therefore potentially sixteen compass roses to a full chart, although most portolan charts show only part of the circle of nodes used to survey the area within the chart, and therefore only a handful of roses.

And then gradually the roses were removed, as cartographers shifted away from portolan charts concentrating solely on coasts to comprehensive maps that strove to incorporate other geophysical data. This shift occurs in the seventeenth century. We can see it happening, in fact, in the two maps in John Selden's *The Closed Sea*. The first map shows Great Britain's position in relation to the seas that surround it. It names countries but does not draw their political boundaries (Fig. 14). The second map is meant to illustrate the King's Chambers (Fig. 15). These were coastal waters across which foreign ships could take innocent passage but could not in any way interfere with ships of other nations. Ships that came within the King's Chambers were deemed to be in British territorial water and therefore subject to British law. James I introduced the concept in 1604 in response to Dutch attacks on Spanish ships within sight of the English coast. It was a unique innovation, the first time a state had ever explicitly defined a zone of maritime sovereignty.

In *The Closed Sea* Selden explains that a committee of twelve men 'very well skilled in Maritim affairs' was convened to determine the boundaries of the King's Chambers. 'These twelv men beginning at the Holy Island, fetch'd a compass round from the North by the East and South to the West.' The Holy Island is Lindisfarne, the site of an ancient monastic community just off the coast in the North Sea, close to the border between England and Scotland. This was the northernmost point of designation. From there the committee drew a series of straight lines from point to point down the east side of England to North Forland (now Margate, at the south edge of the Thames estuary). That stretch, measuring 108 leagues (1 league is 3 nautical miles, or a little over 5½ km), was divided into eleven stages. From North Forland the line continued around to Dover and Folkestone on the English Channel, then ran westwards along the south coast as far as Land's End, for another fifteen stages, measuring a total of 201 leagues. Between this line and the coast lay twenty-six 'very large spaces of Sea', which were grouped into seven King's Chambers. Within these seven Chambers naval or commercial vessels of any nationality could enjoy 'safe riding,' 'safe passage' and 'equal protection', and they were forbidden from plundering each other. James ordered a map of the Chambers to be engraved and distributed so that this new arrangement would be known to all. Unfortunately, the map of the King's Chambers in Selden's book doesn't mark the lines that delineated the chambers. The viewer is left to fill them in for himself.

The two maps look similar enough until you examine them side by side. Then you see that they have almost nothing in common except the wavy shading used to indicate water. As samples of European mapmaking, they point in opposite directions. The map of the King's Chambers points back to portolan charts. It displays three compass roses and hides a fourth behind the distance scale at the bottom right, implying a larger circle of eight nodes. From each of these roses radiate thirty-two rhumb lines in the directions pointed by the eight winds, the eight half-winds and the sixteen quarter-winds. By relying on a web of such lines, the mapmaker had a dense framework within which he could triangulate the exact position of every place on his map, or so he hoped. Another convention of the portolan chart was to name only coastal places, since the purpose of the chart was to guide navigation, and to turn the labels 90° to the coastline so as to avoid obscuring the coastal outline. The map of the King's Chambers innocently reproduces these vestiges of the old portolan tradition.

By contrast, Selden's other map of the seas around England looks away from the portolan tradition: the rhumb lines are gone and the place-names are aligned horizontally. North is labelled at the top of the map (as are the other three cardinal directions along their respective sides) rather than being indicated by a fleur-de-lis at the north point of a compass rose. There is no compass rose. This is no longer a mariner's chart. It summarises, it generalises; it doesn't show the viewer how to get anywhere or, especially, how to follow the line of the coast. The coast has ceded importance to the land.

———————

What, then, is the compass rose doing on the Selden map? Was it an attempt to imitate a conspicuous feature of European maps? It is certainly possible. The Selden cartographer was operating within a zone in which Europeans had been sailing for almost a century. He would have seen their ships and he would have seen their crews, so it is likely that he saw their maps. What cartographer would not want to examine maps that came from outside his own tradition? Is the compass, then, evidence that he was simply copying what he saw?

The foot ruler beneath the rose seems to be an even stronger piece of evidence of borrowing. A distance scale was a strict requirement of European charts. Often it is depicted literally as a ruler over which a pair of calipers or dividers ('compass' in the other, now archaic sense of the word) has been opened. Ruler and calipers were the mapmaker's signature tools, their presence signifying that careful measuring had been done. Together they were the trademark of scientific accuracy. Elizabethan poets picked up the image of the compass/calipers as a symbol of constancy. This is how Ben Jonson uses it when he addresses John Selden in a poem, praising his polymath friend as someone who is

Ever at home, yet have all countries seen:
And like a compass, keeping one foot still
Upon your centre, do your circle fill
Of general knowledge; watched men, manners too,
Heard what times past have said, seen what ours do.

If we are right in supposing that the Selden cartographer drew visual inspiration from European design, this could explain why he extended two rhumb lines 15° to each side of 180° (*bing* and *ding*) downwards from the magnetic compass, producing a triangle that imitates a pair of calipers opened above a ruler.

Stephen Davies, a researcher at the Hong Kong Maritime Museum, has conjectured that the ruler was not purely decorative but was in fact used to determine distances on the map. He proposes that each of the hundred ticks (*fen* or tenths of an 'inch', of which there are ten in a Chinese 'foot') along the ruler's edge marks the distance a ship covered in one watch, of which there were ten in a twenty-four-hour day. If each tenth of an inch marks a tenth of a day, then an inch equals the distance covered in twenty-four hours. Using my estimated speed of 6¼ knots, a day's sail amounted to 150 nautical miles. As a Chinese ruler has ten inches, the length of the ruler on the map should represent 1,500 nautical miles. Davies concedes that 'precise evidence on the map' is lacking; true enough. In fact, the evidence lies elsewhere, and I shall return to it in the final chapter of this book.

Regardless of whether these calculations work, I am cautiously certain that the ruler and compass are on the map as tell-tale signs that the Selden cartographer had seen European sea charts and realised that he could borrow them to excellent effect. They are not purely decorative touches aping the look of a European chart, however, for by transferring this technology to his own production, the Selden cartographer then did what no European cartographer had ever attempted. Rather like John Selden who could imagine lines of jurisdiction over water, he drew the routes of merchant ships over the moving face of the sea.

6
Sailing from China

'Once you are through the harbour entrance, the spray from the whitecaps fills the air and the surging waves leap as high as the Milky Way. No longer can you track the bluffs along the coast. No longer can you note the villages as you pass through them or count off the post stations stage by stage. The senior officers ply the oars and raise the sails, cutting a path through the flood of waves with only the compass needle to show them the way. They rely on its readings to forge ahead, sometimes letting the needle stay on one of the main bearings, sometimes letting it point in a direction between them.'

This is how Zhang Xie describes the exhilaration of launching a junk out of Moon Harbour and heading out onto the open sea. The coast falls away, and with it any fixed point that can tell the pilot where he is going. Now he has only his compasses to show him which way he is going and where he might be. This is anything but stale academic language. Zhang is obviously enthralled with the adventure of seafaring and is determined that the reader should draw the same charge from the thrill of going to sea. Some people have the notion that Chinese were not seafarers and disapproved of compatriots who went to sea, but Zhang clearly is not one of them. I would dearly like to know, but probably never will, whether the author of the *Study of the Eastern and Western Seas* ever got a taste of blue-water sailing. I hope he did.

Even if Zhang did not himself sail through the entrance into Moon Harbour and out into the Taiwan Strait, he knew enough to point the way for us. Fortunately, we have the Selden map and the Laud rutter to confirm and in many places supplement the network of routes that Zhang describes. If each of the three sources mostly confirms the other two, it is not because it had any links with the others. It is simply that the routes they record constituted a stable network that had long been in use by generations of Chinese mariners. Even though we come so many centuries after them, having all three sources makes it easy for us to acquire fairly complete knowledge of the maritime system of which the Selden map is the first fully visualised version (Fig. 1).

There is one small point of disagreement between two of the sources, about which the third equivocates, and that is where they start. Zhang Xie starts unequivocally in Moon Harbour, the port for the prefectural city of Zhangzhou at the southern end of Fujian province, as we have already noted. His account of the departure route is precise and unmistakable: You go out on an ebb tide to Gui Island, which you will recognise with its lighthouse and its shrine to Mazu, for whom the pilots kept their lamps lit. Another half-tide gets you to the coastguard station on Xiamen, which in the nineteenth century would replace Moon Harbour as Fujian's treaty port. Keep going, and in less than five hours you reach the Duster Islands. These form a natural breakwater protecting the outer estuary of Nine Dragon River flowing down from Zhangzhou to the coast. 'Clear the Dusters,' Zhang writes, 'and this is where the Eastern and Western Sea routes divide.'

The Laud rutter does not launch its network of routes from Moon Harbour. Instead it starts at the harbour for Quanzhou, the next prefectural city, 100 kilometres to the north. 'Embark through Five Tiger Gate', the anonymous author explains,

> then take an *yizhen* [115°] bearing as far as Officials' Seawall. After passing three shoals that will surface on the port side at half-tide, put your boat on a *bingwu* [175°] bearing until you have cleared East Sandhill on your starboard side. Once you are sounding six or seven fathoms, set your course on an *yi* [105°] bearing for three watches and you reach Clearwater Islet.

From there you are on open water.

The difference between the two take-off points – Gui Island and the Dusters versus Five Tiger Mouth and Clearwater Islet – reflects the passage of time. The fourteenth century was Quanzhou's heyday as China's main maritime entrepot – that is, until the new Ming dynasty shut down foreign trade in 1374. The only map from that century that shows a sea route leading from China towards the Indian Ocean starts there. Quanzhou regained that reputation in 1403, when the newly enthroned Yongle emperor reinstituted foreign trade and ordered his eunuch Zheng He to lead diplomatic voyages to all states owing tribute to his dynasty, and so the Laud rutter starts there too. During the sixteenth century, however, Quanzhou was eclipsed by its upstart rival, Zhangzhou. Not completely, of course, for this is the home port of the Li brothers in Hirado who served as compradors for the East India Company. But Zhang Xie came from Zhangzhou, and, as he sees the world, Moon Harbour anchored the maritime network that stretched away from China.

Zhang admits that he is using 'the old name' when he calls it Moon Harbour. When the harbour was re-opened for legal maritime trade in 1567, it was given the new, politically correct name of Haicheng, Sea at Rest. On top of this, by Zhang's day the harbour no longer had its distinctive crescent shape. That had been obliterated in the shifting matrix of channels and sand bars choking the mouth of Nine Dragon River. The shallows impeded the bigger junks from getting to shore, which meant offloading cargo onto lighters. Whatever disadvantage that amounted to was more than offset by Moon Harbour's political advantage: it had a customs house. Collecting tax was a racket that spilled money in all directions, most conspicuously to the imperial household itself, but it gave the harbour imperial clearance for foreign trade. It was the edge Moon Harbour needed to go from being a tough seaport where you wouldn't walk alone in the daylight, let alone at night, to being the alpha and omega of an entire network of ports through which Chinese goods were exported to South-East Asia, Europe and the Americas.

The prevaricator among our sources is the Selden map. Look closely at the route lines running in the Taiwan Strait. You will see that they jut in and converge on a node. The right-hand line, which connects to the route running north-east up the coast, is marked *shenmao*, 85°; the left-hand line, which leads to the route heading south-west, is labeled *dingwei*, 205°. Between these two is a third line, which hooks in the Taiwan Strait and then heads to Manila on a constant bearing of *bing*, 165°. None of these lines is drawn so that it touches the coast. All three routes connect offshore, and a dot marks the spot: this is the point from which Ming China's maritime trading network radiated out into the world, the starting point of the Selden map (Fig. 16). Moon Harbour is not labelled, nor is Quanzhou's port, and indeed the small scale of the map rules out including either. The cartographer thus avoids giving either Zhangzhou or Quanzhou priority. In a sense, it doesn't matter. At this scale, both could be used as the port of embarkation. The map thus accords with either the Laud rutter or Zhang Xie's book: no reason to choose between them.

Just as important as that starting dot for understanding the Selden map are the three paths of exit into the Taiwan Strait, for each of these is the initial strand of three distinct webs of routes going in those three directions. This is not easy to detect when first glancing at the map, but scholar Zhang makes the tripartite division of routes crystal clear in his book. In fact, they furnish him with the system by which he has organised his route data. Ming Chinese knew what we call the South China Sea as Nanyang, Southern Sea. With that term taken, Zhang still had the three other cardinal directions to work with, and these he uses to distinguish three different zones of travel from Ming China's south-east coast. Two of them are announced in the title of his book, *Dong xi yang kao*.

Dongyang, the Eastern Sea, denoted the chain of routes that headed east to the Philippines and then turned south to the Spice Islands. Xiyang, the Western Sea, was the sequence of routes that ran south and branched out to the ports of South-East Asia. In practice, the Eastern and Western Sea routes met up in Java, at the bottom end of the great circle they traced around the South China Sea. That leaves Beiyang, the Northern Sea, the routes connecting coastal China northwards to Japan. Zhang Xie shows less interest in the Northern Sea route, so instead of dignifying it as an equal member of Ming China's navigational system alongside the Eastern and Western Seas, he relegates the routes to Japan as an appendix at the end of the book. Our tour of the Selden map starts there.

The Selden map shows the Northern Sea route heading away from China on a *shenmao* bearing of 85°. It then divides into two routes. One runs directly to Kyushu: a straight shot on a *genyin* bearing of 55° that ends at the Goto Archipelago, a string of five larger islands and dozens of smaller ones off the west coast of Kyushu, at the south end of Japan. This was the most direct route linking China to Japan.

A more convoluted route was also available that took ships out to the Ryukyu Islands, a chain of islands of which the largest and the best known is Okinawa. Ryukyu was an independent kingdom that sent tribute to the Ming, although in fact it was under the domination of Japanese lords based in Kyushu. This route heads away from the Fujian coast in a series of six segments that starts on a *chen* (120°) bearing until it fluctuates between *yimao* (95°) and *mao* (90°) as the route picks its way through the southern end of the chain. The cartographer has drawn the islands without any apparent knowledge of their disposition. The Senkaku or Diaoyu Islands, over which China and Japan are currently in noisy public dispute, may be among the lumps of rock that dot the East China Sea between Taiwan and Okinawa, but it would be facetious to suggest that the Selden map can be used to justify anyone's claim to anything in this part of the ocean. The route then runs due north on a *zi* (360°) bearing and then shifts to *guichou* (25°). As it approaches a cluster of islands labelled Yegu Passage, the cartographer inserts a note warning about the fast-moving eastward current (now known as the Kuroshio

Current). The course is then reset on a *yin* bearing (60°), turns 5° to *genyin* (55°) until it comes to the southern end of Japan, then zigzags up the east coast of Kyushu to terminate at a port labelled Hyogo. Adverse weather stranded Richard Cocks in Hyogo on Christmas day 1618. In his diary entry for this incident he notes that he was half a day beyond the outer bar of Osaka harbour. This enables us to identify Hyogo as today's Kobe, one of Japan's main sea terminals.

You would have no idea that this is Kobe by its location on the Selden map. Indeed, you would have no idea what any place in Japan is, for the shape bears no resemblance to Japan as it is or as it appears on any known map, Chinese, Japanese or otherwise. In addition, the place-names are completely bizarre at first glance. The first city south of Hyogo/Kobe is not difficult to decipher. The label reads 'King of Yalima': this is the domain belonging to the daimyo, or lord, of Arima on the east coast of Kyushu. (Cocks met this lord when he visited Hirado in 1621, presenting him with a gift of damask from Canton.) Further south from Arima is a place labelled Shashen wanzi ('Murder Bay'). This one is not too difficult. The character *sha* ('kill') looks quite like a simplified version of the character *gu* ('grain'); make that switch, insert *ga* between the two syllables, open the final n to *na*, and this collection of Chinese syllables sounds out the Japanese place-name Kagoshima – the large bay at the southernmost point of Kyushu. Continuing further around the end of Kyushu, we come upon yet another strange name, Shazima ('Killer Horse'). This is the Selden cartographer's version of Satsuma, the most powerful domain on the island of Kyushu.

Japanese place-names are always written in Chinese characters. Killer Horse and Murder Bay are not how Satsuma and Kagoshima are written, from which we have to deduce that the Selden cartographer had no idea how to write these place-names. Instead he has transcribed how he has heard them pronounced, and not by a Chinese who knows the proper characters for these names. This hypothesis gets stretched further by the next example on Kyushu, Nagasaki, the only port that the Tokugawa government allowed to remain open, and then only to a handful of Dutchmen, after it closed the country in 1641. The label reads Longzishaji. This mouthful is more than just a heroic attempt to render Nagasaki, for Europeans, adapting Portuguese usage, knew it as Langasaque. Longzishaji clearly derives from the Portuguese rather than the Japanese pronunciation. Some Chinese knew both names for the port, for both Nagasaki (written in its proper Chinese characters) and Longzishaji appear in the Laud rutter. The Selden cartographer didn't. He knew only the European version. We can't conclude from this that he heard it directly from a European, but if he didn't, he heard it from people who were involved in trade with Europeans.

Up the west side of Kyushu beyond Nagasaki lies a place-name that baffled me for the longest time. It reads Yulindao: 'Fish Scale Island'. Rather than get confused by the meaning, we have to work once again from the sound. *Yu* is a highly unstable pronunciation in Chinese as well as being difficult for foreigners to pronounce, so it could be standing in for almost any soft or breathy sound. *Lin* can move directly across both languages, but it can also get mangled within the family of sounds known as alveolar consonants: n, l and r. Fortunately *dao* is easy: it corresponds unfailingly to the Japanese *do*; words such as 'judo' and 'kendo' derive their *do* from this Chinese syllable. 'Fish Scale Island' turns out to be an approximation of another Portuguese place-name, Ferando, which is how Europeans pronounced the name of the port where John Saris opened the East India Company factory, Hirado.

The records of the Company's decade in Japan – the diary of Factor Richard Cocks, the journal of Captain John Saris, the logs and letters of Master Will Adams –provide the detailed information about what it was like to travel the arteries of the Northern Sea, far beyond anything Zhang Xie was able to record. For them, as for Li Dan and the other merchants operating in Hirado and Nagasaki, the Northern Sea route was their lifeline to the Ming economy. Given that access to China was difficult, if not impossible, it was also their conduit to the ports in Annam, Champa (both part of today's Vietnam), Cambodia and Ayutthaya (Thailand), and beyond these to Pattani and Johor down the Malay Peninsula and ultimately to Bantam and Batavia (Jakarta) on Java. Japan-based merchants treated the Northern Sea route as a backward extension of the Western Sea route. Together these routes constituted a sort of maritime freeway along which ships of all nations came and went. The China Captain and the East India Company worked this route as intensively as they could, but it was not easy.

The English used both arteries of the Northern Sea route. When Saris left Japan on the *Clove*, he took the more direct westerly artery. Leaving Japan in December 1613, he 'steered away South-west, edging over for China', as he writes in his journal. 'A stiffe gale and faire weather' blowing from north-north-east provided perfect conditions for a ship sailing on the reciprocal. Seven days later the *Clove* was off the Fujian coast. Near Moon Harbour, three hundred large junks cut across Saris's path on their way out to fish. This striking sight was repeated three days later off the mouth of the Pearl River below Canton. The *Clove* curved on a path south-west 'as the Land trends' and four days later stood off the mouth of the Mekong River. It was an easy run, by far the easiest that any EIC ship would have coming out of Japan.

The first ship Richard Cocks sent down the Northern Sea route was the *Sea Adventure*, a leaky hulk he bought in Nagasaki the following summer. Before despatching it, he assembled whatever cargo he could that autumn so that it had something to trade when it reached its destination of Siam, today's Bangkok. The total value of the cargo was only 700 taels (ounces of silver), so he also stowed 5,000 taels in raw silver to buy commodities there that he could sell for a profit back in Japan. He appointed Will Adams as captain, engaged Edmund Sayers and an Italian named Damien Marin as pilots, and put Richard Wickham in charge of the voyage's commercial operations. In his written instructions to Wickham he advised against selling the cargo too cheaply; better to just bring it back than sell it at a loss. He listed what Wickham should buy in Siam: first aromatic wood, which sold well on the Japan market; after that deer skins, Chinese textiles, dried fish skins (used to cover scabbards and sword handles) and buffalo horns, which were currently imported from the Philippines and could perhaps be undersold if acquired elsewhere at a lower price. Cocks gave Wickham the going rates for these commodities in Siam as well as their prices in Japan so that he would know how much to pay. Wickham should in any case not let filling the hold delay his departure too much after the onset of the summer monsoon: better to come back only partly laden rather 'than unadvisedly to adventure the rest'. To these commercial instructions Cocks added two pieces of personal advice: don't get into arguments with the prickly Adams and avoid 'the feminin gender, although the liberty of these partes of the world is overmuch in that kinde'.

Foul weather delayed departure of the *Sea Adventure* until 17 December. According to the Selden map, Adams would have set out on the reciprocal of *genyin* (55°) (that is, *qianxu*, 305°) and sailed straight along the coast of China. Three days out he found that the *Sea Adventure* leaked so badly that he changed to the easterly artery and headed on the reciprocal of *guichou* (25°) (that is, *renhai*, 335°) and headed for the Ryukyu Islands, reaching Okinawa on 27 December. Four days later he grounded the junk to perform the necessary repairs. This was a slow process: laying the mast, emptying the ballast, washing the hull, then inspecting for leaks. The only serious leaks were around the nail holes, so he had his crew re-caulk the planks on the hull. The work was stalled, however, when he discovered that the local lime had been adulterated. Meanwhile Adams had to placate the local officials, who knew they could not offend the Japanese (whom the Ryukyuans understood to be the patrons of the English) but who could not risk offending China (a Ming delegation was expected on Okinawa shortly and would certainly object to Europeans in the islands). He also had to defuse conflicts between his Japanese crew and the Japanese merchants and their servants, numbering over twenty, who were travelling as paying passengers. Adams stepped in to prevent an armed battle between the sailors and the merchants, but tensions grew to such a pitch through March that the chief of the merchants ended up murdering the crew's ringleader. Then there was the weather, which refused to cooperate. When it was fair, the wind blew at their faces from the south-west. As soon as the wind swung round to the north-east, the weather turned foul again. The crew descended into disorder, theft and rape. Wickham was fighting with Adams, as Cocks feared he might, and the Ryukyuan officials were desperate to get the *Sea Adventure* off their beach. By the time it was possible to launch the ship, on 21 May 1615, it was too late to sail to Siam and hope to return on the monsoon winds. There was nothing to do but sail back to Hirado.

Adams tried again the following December. The journey began in foul weather, although this time the hull did not leak and the winds were with them. The ship made excellent headway. In seven days the *Sea Adventure* stood off Quanzhou, and six days after that it was off the coast of Vietnam. Three weeks later Adams and Sayers were in Siam. Business was good. Sayers was able to acquire so much for the Hirado factory that he had to hire a second vessel in Bangkok to ferry it all back. Will Adams set off on the *Sea Adventure* on 5 June 1616 just in time to ride the north-westerly monsoon, and reached Hirado on 22 July.

No such luck for Sayers. He left Bangkok only one day later, but the winds changed and it took him twelve days to clear the bar at the anchorage beyond Bangkok. By the time he was out on the South China Sea the wind had died. The junk got as far as the coast of Fujian by the end of the month and then was unable to go any further. By then the winds had turned and blew in their faces out of the north-east. The captain shifted the junk helplessly from one offshore anchorage to another, unable to make any headway. The Buddhists threw little Siamese pagodas overboard on 9 August to seek divine help. The Chinese captain sacrificed to the Empress of Heaven three days later. The Japanese Christians held their own ceremony. The wind finally swung round to the south on 22 August, but by this time over half of the crew was sick with scurvy. The men started to die, including the captain. Sayers finally got the junk back to Japan, although when the accounts were cleared, Cocks discovered that two of the Chinese officers had fiddled the books and cheated the Company of much of the profit it might have realised from the trip.

Adams bought Sayers's junk on his own account, renamed it the *Gift of God* and took it on a successful winter run down to Vietnam. The *Sea Adventure* also sailed that winter under a Japanese captain and two English pilots but faced even worse luck than it had suffered the previous year, losing thirty-four crewmen before it could get back to Hirado. Adams sold the *Gift of God* at the end of 1617, but the *Sea Adventure* made one last voyage. Once again it leaked so badly that it had to shift the route to the Ryukyus. Almost a year later the *Sea Adventure* was able to set sail. It made Siam on the last day of December, but was deemed unfit for further service and junked, to use that word in its other sense. Two more junks sailing for the EIC would run down the Northern Sea route, one as far as Tonkin (Hanoi), where it turned a modest profit, the other as far as the Mekong Delta, where it was blown back by winds and earned the Company nothing.

With no further sailings over the next four years the EIC pulled Cocks and his men out of Hirado and closed the factory. Cocks was reprimanded for losing the Company a great deal of money and ordered back to London. He died in agony en route and never made it home. The real problem wasn't Richard Cocks. It wasn't the difficult monsoon winds, for all the havoc they could wreak when the timing was wrong. It was the conditions of trade: too much competition, too little access to Chinese markets and too much extortion from officials every step of the way.

———————

For those who could not trade directly into the Ming dynasty, which in this era was just about everyone, the Northern Sea route was not sustainable on its own. The only way to trade profitably was to link up, as Will Adams and Edmund Sayers repeatedly did, with the Western Sea route. On the Selden map that route starts out by leaving Zhangzhou/Quanzhou on a bearing of *dingwei* (205°), then once in the Taiwan Strait turning to *kunshen* (235°). The Selden map shows the Western Sea route splitting many times as it goes south, sometimes into two routes, sometimes into three, at each fork sending out tendrils around the South China Sea. The first fork is at Qizhou, the Seven Islands, off Hainan. Here one route splits off on a long *gengyou* bearing (265°) that then turns to *qianhai* (325°) and runs into the port serving Tonkin (now Hanoi). The main route continues south between the Paracel Islands and Vietnam, then divides four ways off the south-east coast of Vietnam: one strand straight to Batavia, one that makes a long arc south and then east around Borneo, a third to the mouth of the Singapore Strait, and a fourth to Pattani, half-way down the Malay Peninsula. Each of these strands linked China to yet another Chinese sojourning community based in ports around the South China Sea.

One of the busiest points on the Selden map is the Singapore Strait, at the south end of the Malay Peninsula. This was the location of the state of Johor, famously the site of van Heemskerck's seizure of the *Santa Catarina,* which led to the debate between de Groot and Selden. Routes from Johor go in every direction. One heads east across the South China Sea to Brunei. One heads south-east (*xunsi*, 145°) around the south end of Borneo towards the Moluccas, with a later branch veering off down to Surabaya at the east end of Java. Another takes a *dingwu* bearing (185°) down to Palembang on the east coast of Sumatra, and from there goes to Bantam and Batavia on Java. One more route passes through the

1. The Selden map.

2. Extract from the codicil of John Selden's will, dated 11 June 1653, in which he directs his executors that they should donate his 'Mapp of China made there fairly' to the University of Oxford, though the attested copy of the will fails actually to name the university.

3. John Selden as a younger man, perhaps in his thirties while he was still enjoying the company of Ben Jonson and John Donne but gaining a reputation as Britain's foremost expert on constitutional law.

4. The poet Ben Jonson at middle age, about the time John Selden was called before his patron, James I. Jonson was Selden's closest friend in his early years in London and his protector when he could afford to be.

5. Huig de Groot (Grotius) in 1608 at the age of twenty-five. De Groot had already written *The Open Sea* and would publish it anonymously the following year. He and Selden became intellectual sparring partners in the course of defending the claims of their competing sponsors; they also became each other's greatest admirers, and between them laid the foundations of modern international law.

IOHN   SELDEN.

6. John Selden later in life. He has shed the sensitive uncertainty displayed in his earlier portrait for a tougher, more cynical but not unkindly demeanour.

THOMAS HYDE S.T.D. PROTOBIBLIOTHECARIVS XI DEC MDCLXV
OFFICHIM SPONTE DEPOSUIT IX APR. MDCCI

7. Thomas Hyde was the first and only scholar of Oriental languages to be appointed Keeper of the Bodleian Library, in 1665. This portrait, by an unknown painter of modest skill, was executed shortly before Hyde's retirement in 1701. Notice the writing on the scroll in his right hand: a few characters in Chinese. The painting hangs in the Schools Quadrangle of the library in the room that currently serves as the gift shop.

The Royal Gift of Healing

R. White sculp.

8. 'The Royal Gift of Healing'. This popular print shows Charles II healing victims of scrofula by a public ritual known as 'touching for the King's Evil'. It was a common practice of European monarchs to lay healing hands on people suffering from this form of tuberculosis, known then as the King's Evil. The last Stuart king to do so was James II, on the occasion of his final visit to Oxford in 1687.

9. The royal portraitist Godfrey Kneller painted this portrait of Michael Shen,
also known as Shen Fuzong, in 1687 at the request of James II. The painting,
popularly called *The Chinese Convert*, commemorates the visit of the first Chinese
to Great Britain. Kneller's depiction consciously celebrates Shen's conversion
to Catholicism. Shen posed for the portrait in robes brought from China.

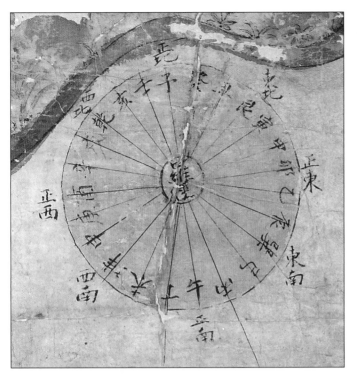

10. The compass rose on the Selden map.

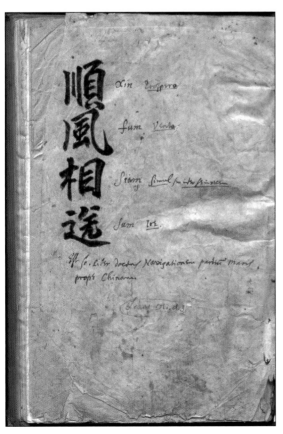

11. The cover of a manuscript Chinese seafaring manual in the Bodleian Library. It is known as the Laud rutter in honour of Archbishop William Laud, who donated it as part of a larger gift of Oriental manuscripts in 1638. The title in Chinese, *Shunfeng xiangsong* ('Despatched on Following Winds') may have been written by Michael Shen during his visit to Oxford in 1687. What is certain is that Shen wrote the accompanying pronunciations in Roman script for the four characters: Xin Fum Siam (corrected to Siang) Sum. The translation into Latin was done at the same time by the librarian Thomas Hyde.

12. A Chinese compass with wooden cover, which John Selden donated to the Bodleian Library. The needle is dry-mounted in the centre, around which are marked the eight trigrams of the *Book of Changes* in the innermost ring and the twenty-four main compass directions in the fourth concentric ring. On long-term loan to the Ashmolean Museum, the compass is now on display in Oxford's Museum of the History of Science.

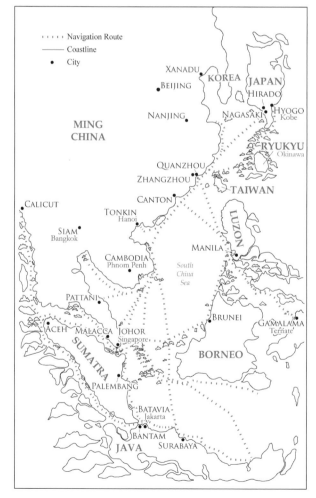

13. The Selden map transcribed, showing place names and routes. Place names in capital letters are those in use at the time; their modern equivalents appear beneath them in smaller font.

14. Map of England and part of Scotland which Selden included in *The Closed Sea* to illustrate the 'King's Chambers', zones of jurisdiction over Britain's coastal waters. The design of the map reflects an older style of mapmaking used on portolan charts, which were drawn on a web of rhumb lines radiating from multiple compass roses, and named only points on the coast, which always appear at right angles to the coast.

15. Map of Great Britain and its surrounding seas, the second of the two maps in Selden's *The Closed Sea*. The design of this map has abandoned the portolan tradition of compass roses and rhumb lines in favour of highlighting places inland. Printing place names horizontal to the viewer's perspective marks another break with portolan practice.

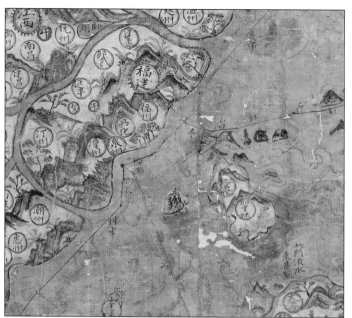

16. In this excerpt from the Selden map, the dot in the Taiwan Strait off the southern coast of Fujian province marks the starting point of the network of routes that extends across the map. The two closest prefectural capitals, Zhangzhou and Quanzhou, were the two main ports from which Chinese navigators sailed abroad during the Ming dynasty.

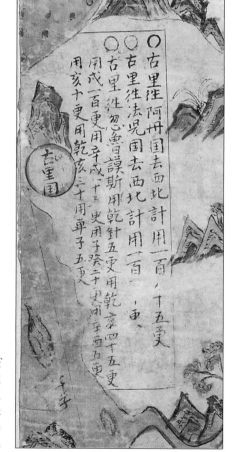

17. The Calicut cartouche on the right side of the Selden map provides the distances and directions for sailing from Calicut (circled in red), a major trading centre on the west coast of India, to the three main ports on the Arabian coast: Aden, Djofar and Hormuz.

18. The English editor and writer, Samuel Purchas, as depicted on the title page of his popular four-volume collection of travellers' tales, *Purchas his Pilgrimes*. This book presented most English readers with their first maps of China, which are reproduced in the two illustrations that follow. Speed and Selden were great friends until they fell out after 1617 over Speed's careless use of Selden's research in a previous book.

19. Purchas took this map directly from the Dutch map publisher, Jodocus Hondius, who copied it in turn from the great Abraham Ortelius in 1584. All three distinctively turned China so that north was to the right. In Purchas's view, this was a flawed European misrepresentation of China. He included the Hondius map in his compendium for the purpose of contrasting it with the Chinese map that follows.

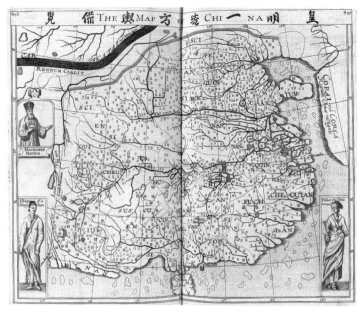

20. John Saris acquired the original for this map from a Chinese merchant in Bantam (Java) in forfeiture for unpaid debts to the East India Company. Saris gave the map to the editor and publisher Richard Hakluyt, from whose estate it passed to Samuel Purchas. Purchas included it as the correct representation of China, as it was copied directly from a Chinese original. That original has long since been lost.

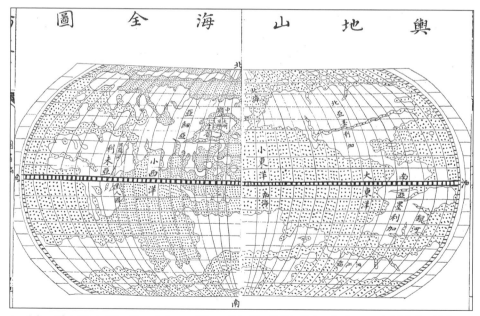

21. *A Complete Map of the Mountains and Seas of the Earth* (*Yudi shanhai quantu*). This is a Chinese copy of a map that the Italian missionary Matteo Ricci printed in 1584 based on a map designed by Abraham Ortelius in the pseudocylindrcal projection. A distinguishing trait is its depiction of a water passage across North America, a passage that northern Europeans hoped to find in order to shorten the journey to China and sidestep their Portuguese and Spanish competitors.

22. John Speed was a tailor who trained himself to become a cartographer and map publisher. He designed and engraved this map of Asia in 1626 in the course of producing his great world atlas, *Prospects of the Most Famous Parts of the World*, which appeared the following year. Speed was eighty-five when the atlas was published, and died two years later.

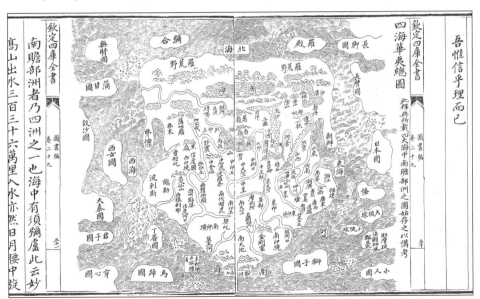

23. *General Map of Chinese and Barbarians within the Four Seas* (*Sihai huayi zongtu*), as published in the 1613 edition of Zhang Huang's *Tushu bian* ('Documentarium'). The map draws on Buddhist cartography to fashion an image of what we call the Eurasian continent. China occupies the south-western quarter of this continent, its northern border indicated by a crenellated Great Wall. Much of the map beyond China is imagined and many of the place names are mythical.

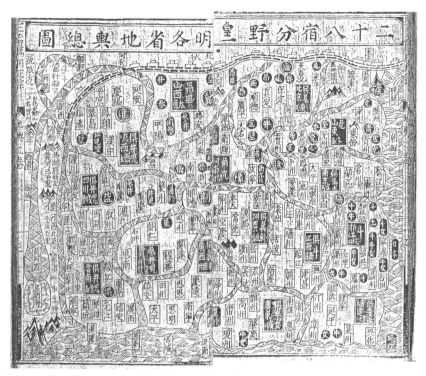

24. *General Topographical Map by Province of the Divisions and Correspondences of the Twenty-Eight Lunar Mansions* depicts China in relation to the constellations that were thought to govern the dispositions of various parts of the country. This map, published in the 1599 edition of Yu Xiangdou's bestselling household almanac, *Wanyong zhengzong* ('Complete Source for a Myriad Practical Uses'), was widely recopied in the popular press. Its content is almost identical to the China portion of the Selden map, and may have served as a source.

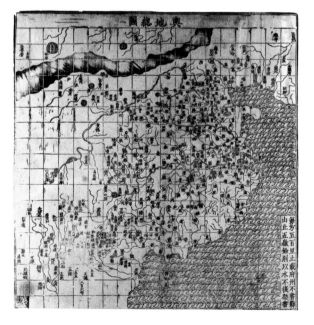

25. This map of China appears as the first plate in Luo Hongxian's atlas, *Guang yutu* ('Enlarged Terrestrial Atlas'), first published in 1555. Luo's atlas was regarded as the standard when the first Jesuit missionaries arrived in China. They sent copies back to Europe, where cartographers took up features of Luo's maps and incorporated them into their own.

26. The Selden map geo-referenced. In the course of aligning the Selden map with its GIS coordinates, we discovered that the best fit between what the Selden cartographer drew and the actual shape of East Asia (in red) resulted when we broke the map into sections and allowed the landforms to move apart around from the South China Sea. Once that adjustment was made, the map revealed a degree of accuracy that is astonishing for its time.

27. Prince Giolo, as he was briefly known in England, was a native of Miangis Island southeast of the Philippines. He was captured by Muslim slave traders and later brought to England in 1692. His elaborate full-body tattoos made him a popular spectacle from the palace down to Fleet Street taverns. After he died of smallpox that autumn in Oxford, he was flayed for preservation. His skin hung for some years beside the Selden map on the stairway leading to the Bodleian Library.

28. Freebooter and privateer, William Dampier was less successful in commanding profitable naval expeditions than he was in writing about them. Thomas Murray painted his portrait about the time Dampier published his bestselling *A New Voyage Round the World*, which appeared in its first edition in 1697. While he was in the Philippines, Dampier acquired part ownership of Giolo, whom he brought with him back to England in 1692.

Singapore Strait and heads north-west (*qian*, 315°) up the Malacca Strait to the port of Malacca (now Melaka), then continues on up to Aceh at the north end of Sumatra. The final leg in the Western Sea network runs north out of Aceh into the Indian Ocean.

Moon harbour was also the place to embark on the third sea lane out of China, the Eastern Sea route. Zhang Xie describes the route heading out on a *bing* bearing at 165°. The Selden map does not label the direction of the first short segment, but labels the main line of navigation that slips past Taiwan and runs directly to the Philippines as a *xun* (135°) bearing. The cartographer takes no interest in Taiwan, for the simple reason that none of the early history for which it is famous – Dutch trading stations, immigration from Fujian, the rise of the Zheng family's Eastern Calm dynasty and its suppression by the Qing dynasty – had yet happened. At this point Taiwan was of no interest. The Eastern Sea route instead makes a beeline on a *bing* course (165°) for Manila. The string of seven named ports dotting the north-west coast of Luzon, the largest of the Philippine Islands, suggests considerable geographical knowledge of the area. Before the *bing* course gets to Manila, it joins with a *xunsi* course (145°) coming from Canton, or more precisely from the Portuguese colony of Macao downstream from Canton at the mouth of the Pearl River. The two routes for sailing from China to the Philippines flow together onto a *bing* bearing (165°) and run straight into the harbour at Manila.

Manila was the entrepot where the Spanish empire of conquest met the Ming empire of trade, where cargoes of Chinese merchandise were swapped for chests of American silver. There is no indication on the Selden map that Manila is a European city, but the naming of the Parian, the ghetto to which the Spanish consigned Chinese residents, across the river from Manila indicates an awareness that this was an entrepot where Chinese went to trade but had to live apart. A place-name just to the south of Manila marks the entrance to what appears to be Verde Island Passage, which runs along the southern side of Luzon. Follow that passage eastwards through the San Bernardino Strait and you reach the Pacific Ocean. The Selden cartographer draws no line through the strait – he must have lacked the navigational data to do so – but he does insert a short note at the point where the passage opens out to the ocean: 'Shapeshifting foreigners go via this anchorage to and from Luzon.' This was the point at which the Manila galleons headed out into open sea to make the trans-Pacific crossing for Acapulco. These 'shapeshifting foreigners' have to be Spaniards. The term 'shapeshifter' (*huaren*) is one I had never encountered in a Ming source before. The place to find it is in a Han-dynasty collection of ancient myths and tales known as *The Writings of Master Lie*, which includes a story about people who come to China from a 'country far to the west' and whose powers to transform themselves defy the physical limits of this world. When strange-looking Europeans first arrived in the waters around the Ming dynasty early in the sixteenth century and said they were from the far west, people reached back into those old tales to find them a name.* The cartographer did not know enough about sailing through the archipelago to plot the route, but he did know that this was the exit point for ships heading to the Americas, and the entrance for the galleons that arrived annually laden with Peruvian silver to buy Chinese goods in Manila. We know it now as the channel connecting the Chinese and European economies in the seventeenth century. Did he? Possibly.

Outside the mouth of Verde Island Passage, the routes divide. The western strand heads straight to Brunei, on the north-west coast of Borneo. From Brunei it veers onto a *shengeng* bearing (250°) to a pair of islands off the west corner of Borneo, then turns due west and shoots across the bottom of the South China Sea to the Malay Peninsula. For Zhang Xie, Brunei marks the end of the Eastern Sea route. The Selden cartographer, however, turns Brunei into one of several points where the Eastern and Western Sea routes connect.

---

* This term 'shapeshifting foreigners' (*huaren fan*) appears again in a rectangular label at the top of the map above and just to the right of the compass rose. The writing is badly worn, but my best guess is that it reads: 'shapeshifting foreigners reside beyond here.' The location at the eastern end of the Gobi Desert suggests Jurchens (ancestors of the Manchus who would conquer China in 1644) or possibly Mongols, although there is another label to the left of the compass that reads, 'northern Tartars reside here', a term widely applied to steppe peoples. Did he regard the Jurchens/Manchus then banging on the gates of the realm as being of the same category as Europeans? We have no reason to think so, but then we have no reason to think anything else.

The other strand of the Eastern Sea route south of Manila takes a south-easterly course. Without either compass bearings or identifiable place-names, it is difficult at first to figure out where the cartographer is sending his ships. Nor do the shapes of the islands betray the reality they might be standing in for. The only name we can identify with confidence among the scattering of east-tending islands south of Luzon is Sulu. The Sulu Archipelago is still called that today, although it sits on the same degree of longitude as Manila, due south rather than way off to the east, which is where the Selden map has it. This is clearly not a part of the world with which our cartographer is familiar. What he knows is that there is a nautical route connecting Manila to Sulu, not where Sulu itself is located. In fact, his line doesn't quite get to Sulu but halts at the next port to the west, as though uncertain exactly how these places connect. His pen has wandered into terra incognita.

Still the line continues on to the oddest route on the map: a zigzag leading from the southern end of the Philippines to a place called Wanlaogao. Is there precision to this line, or is it a random rendering of a connection that he knows is there but doesn't have the compass bearings for and can't really locate? The latter, it seems. Wanlaogao was for me the toughest puzzle on the map. *Gao* means 'lofty', and *wanlao* means 'as old as ten thousand years'. The name suggests an old mountain, but that doesn't much help. Insular South-East Asia is a volcanic zone; the mountains here are too numerous to count. Fortunately, Wanlaogao makes an appearance in Zhang Xie's *Study of the Eastern and Western Seas*. There it is listed as one of two topographical features of an even more mysteriously named place called Meiluoju. Zhang says the place is also known as Miliuhe, although that doesn't help in identifying this spot either. To explain Wanlaogao, this is the story Zhang gives, more or less in his own words.

The country had previously been invaded by a people known as the Franks. This was a generic term Chinese used for Europeans, borrowed originally from Arabic, although in this case it means Spaniards. The indigenous people surrendered, and the Franks extended an amnesty to the chief, ordering him to continue administering the country as before but to deliver to them a fixed quota of cloves annually. The Franks did not impose direct military rule but left the country to defend itself. Then another group of marauders crossed the ocean, the Red Hairs – the Dutch. Once they arrived in the region, no place was secure. Their ships made a surprise attack on the town and captured the chief.

'If you serve us well,' the Red Hairs told him, 'then we will make you the master and drive out the White Throats.' Zhang then explains that Franks have white throats, hence the name. The captured chief had no choice but to agree to their terms. As soon as the White Throats got wind of this, they came back and accused him of double-dealing.

'You traitor of a slave,' they railed. 'We always regretted not putting you to the sword, and now you rebel!' And so the White Throats attacked.

At this point Zhang weaves in another story. He writes that the king of Luzon, with the baffling name of Old Langleishi Bilixi, levied Chinese working in Manila to man the ship in which he put to sea to take back Meiluoju. Suffering under his cruel command, the Chinese attacked him in his cabin one night. As Zhang tells the story in greater detail elsewhere in his book, the Chinese ringleader, Pan Hewu, declared to his fellow conscripts, 'If we don't die of treason or hanging or stabbing or whatever, we will certainly die in battle. Better to kill the commander for mistreating us. If we're victorious, we raise the sails and return home. And if we're not, we won't have to wait so long to die.' So they killed him, took over the ship and made for Vietnam. The king's son, who was serving elsewhere in the Philippines, hurried back to ensure his succession as soon as he heard the news rather than pursue the campaign against Meiluoju. Much later, he would raise an even larger force to realise his father's ambition.

This, at least, is how Zhang tells the story. Spanish sources relate this tale from the other side. The king, Old Langleishi Bilixi, was in fact the Spanish governor of the Philippines, Gómez Pérez Damariñas. He was indeed murdered by conscripted Chinese seamen, and his son Luis Pérez did succeed his father as governor, racing back to Manila to prevent the post from falling into the hands of the senior Spanish military commander. The problem with Zhang's account is that Damariñas was murdered on 25 October 1593. Dutch ships were not yet prowling in these waters. Zhang has mixed up his stories, weaving what he has heard about the Spanish and the Dutch at different times into a single story. He goes on to explain that the Dutch went home every year and a new batch arrived the following year to take their place. This made their presence intermittent, and while they were away, the Spanish moved in. As a result, control of the island of Meiluoju swung from the Dutch to the Spanish and back again. The situation was stabilised not by the Europeans but a Chinese trader sojourning on the island. To reduce tensions this merchant, whom Zhang describes as crafty and good at persuasion, got the two European parties to agree to divide the island between them. The line of division would run through a mountain called Wanlaogao. North belonged to the Red Hairs, south to the White Throats.

Zhang has most of the story right and some of it wrong. Gómez Damariñas in 1593 was on his way to the Spice Islands to drive the Portuguese out and take control of the tiny island of Ternate. The mutiny scotched that plan, although it would take two more attempts before the Spanish were able to secure control of Ternate in 1606, carrying its sultan, Said Barakat, back to Manila as a virtual hostage. They enjoyed dominance for only a year, for the Dutch arrived in 1607 to build themselves a fortified base. Zhang is right that the Dutch tried to impose a presence earlier, in 1600. As it happened, the VOC captain who arrived in 1601 to take them off was Jacob van Heemskerck, two years before he seized the *Santa Catarina*.

So is Ternate Wanlaogao? Their identity was clinched when I chanced upon a 1726 Dutch engraving of Ternate. The inset in the upper left-hand corner shows the ground-plan of the fortified Spanish settlement, there named Fort Gamma Lamma. Gamalama is the name of the volcano on which this tiny island sits. Chinese isn't a good language in which to transcribe anything but Chinese. To pronounce Gamalama, southern Chinese would drop the first syllable *ga*, and would nasalise the second syllable as *man*, which in Mandarin would become *wan*. *La* is a weak syllable in Chinese: it would slide to *lao*. The *gao* at the end simply means 'tall', which is what mountains are. And so from the Malay name Gamalama we arrive at Wanlaogao.

Ternate was the furthest outpost of the spice empire as well as the further reach of the Chinese trading empire. As for the alleged line of division, that too is true. The Dutch landed and built a fortified base in 1607 on the other side of Gamalama from the Spanish and divided the island between them until the Spanish withdrew in 1663. The Selden map includes two labels that commemorate this story. One of these – 'Where Red Hairs live' – caught Hyde's attention and got him scribbling 'Hollanders' in the margin. These are the Dutch interlopers. The second label beside it reads, 'Where shapeshifters live'. This, of course, refers to the beleaguered Spanish.

What the Selden map doesn't show is the route connecting Ternate to the rest of the South China Sea network, which John Saris took to get there in 1613. We know that Chinese sailed this way on the strength of Saris having kidnapped the master of a Chinese ship en route, yet the map gives no trace of that link, which would connect the Eastern Sea route out to the Philippines with the Western Sea route down to Java. For Chinese mariners the two routes formed a great circle, but not for the Selden cartographer. Ternate was the outermost point that his knowledge reached. Indeed it barely reached there at all, everything on that side of the map beyond Borneo being a confused swirl of overpainting that hedges what is water and what is land.

If Ternate was at the edge of being out of range for the Selden cartographer, so it was too for John Saris. Having nothing to tempt the Spanish beyond a few pairs of seamen's boots, Saris gave up on the hope of turning the Spice Islands into a free sea and turned north to Japan, plotting a route across open ocean to the east of the Philippines, which no Chinese vessel would have taken and which the Selden map does not mark with a line. Saris sailed beyond the outer edge of the Eastern Sea route to re-enter the system from the far end of the Northern Sea route. That end-run would not provide the asset the Company needed to earn enough to justify staying on in Japan waiting for the China trade to open. Seven years after Saris returned to England with his wealth and pornography, the EIC pulled the plug on the Japan factory and called Richard Cocks home, a place he never reached.

From the circle of routes that the Selden map depicts, there was only one westward exit. You can find it by going to Johor, at the end of the Malay Peninsula, the spot that Selden may have worn away with too much touching. There you will find the final spur of the Western Sea route as it runs up the Malay Strait to Aceh at the north end of Sumatra. The route does not end there. It splits. The western strand curls around the top of Sumatra and runs down the outer side, a zone that no surviving rutter records, although there were ports along that coast. The eastern strand finds a *renzi* bearing (355°) almost due north in the direction of Burma. And then it comes to an abrupt dead-end at Calicut, a port city in Kerala, on the west coast of India.

What is Calicut doing here? The spot marked on the Selden map corresponds roughly to Rangoon. What happened to the Bay of Bengal, to say nothing of the Indian subcontinent? The Selden cartographer is not fazed by the sudden collapse of the maritime world beyond Aceh. He just keeps going, not in space but in words. This is the only place on the map where he inserts a cartouche, a box explaining what is going on (Fig. 17). Here tight to the left-hand margin of the map he provides three striking notes in bullet form. The first reads:

- Calicut to Aden: go north-west for 185 watches.

Aden is in Yemen, on the south coast of Arabia near the mouth of the Red Sea. The cartographer clearly treats the place he has labelled Calicut as Calicut, and is explaining a route that crosses not the Bay of Bengal but the Arabian Sea. For the first time on the map he provides no compass bearing, only the ordinal direction 'north-west'.

The second note is much the same:

- Calicut to Djofar: go north-west 150 watches.

Djofar is further east along the same coast in what is today's Oman.

The final direction is more detailed and, in lieu of an ordinal direction, supplies precise compass bearings:

- Calicut to Hormuz: stay on a *qian* bearing (315°) for 5 watches, *qianhai* (325°) for 45 watches, *xu* (300°) for 100 watches, *xinxu* (295°) for 15 watches, *zigui* (10°) for 20 watches, *xinyou* (275°) for 5 watches, *hai* (330°) for 10 watches, *qianhai* (325°) for 30 watches and *zi* (360°) for 5 watches.

Aden, Djofar and Hormuz were the three great ports of medieval Islamic trade before the intrusion of Europeans in the Indian Ocean, but they were not destinations for Chinese ships. Even in the Yuan dynasty, Chinese sailed no further west than Calicut, where they transferred their cargoes to the ships of Muslim merchants. If the routes described on the cartouche were not for Chinese navigators, who were they for? The answer is that they were not for anyone. They are in fact records of the voyages of the imperial eunuch Zheng He who visited all three in the fifteenth century. What this tells us is that our cartographer was working from a source in which these routes were indicated – and we can find routes very much like these in the Laud rutter, though not in Zhang Xie's *Study of the Eastern and Western Seas*, which came too late to bother with Zheng He. The routes he has drawn across his map, then, illustrate a written text that has disappeared at the same time that it shows the commercial networks of the China seas late in the Ming dynasty. It is a composite of text and experience.

Finding Calicut on the Selden map resolves one of the puzzles left over from our foray into the life of Thomas Hyde. Remember the Chinese writing on the scroll in Thomas Hyde's hand? Go back and take a look. The two prominent characters at the top are *gu* and *li*, 'ancient' and 'reason'. Combine them and you could plausibly come up with the phrase 'ancient principle', but in fact what you have is the Chinese attempt to write the first two syllables of the word Calicut, *Guliguo*. There is nothing on the map to suggest why Hyde should have selected Calicut for his portrait and nothing in his notes that betrays a personal interest, yet it is hard to believe that a scholar of his temperament would have chosen the characters at random. Might this Indian Ocean seaport have represented some sort of far edge of what a European could know about Asia, beyond which lay realms of knowledge beyond access? Did Hyde perhaps think of Calicut as a meeting point between east and west?. Hyde has inscribed the Chinese rendering of Calicut on his portrait. He didn't annotate Guli on the map, nor does the term appear anywhere in his notes. Clearly he was paying attention when Michael Shen showed him Calicut on their map – but to what, I wonder?

7
Heaven is Round, Earth is Square

In 1625 Samuel Purchas gave English readers their first maps of China. Today his name means nothing to most people, unless you happen to be an aficionado of seventeenth-century travel literature. In his own day, though, his collections of travellers' tales were everyone's favourite recreational reading, and his name became one of the best-known in the world of popular publishing. The books bearing his name were still going strong in 1798, when Samuel Taylor Coleridge famously fell asleep over a passage on page 472 of *Purchas his Pilgrimage* – 'In Xamdu did Cublai Can build a stately Palace, encompassing

sixteene miles of plaine ground with a wall, ... and in the middest thereof a sumptuous house of pleasure' – and awoke to write one of the most famous poems in the English language:

> In Xanadu did Kubla Khan
> A stately pleasure-dome decree:
> Where Alph, the sacred river, ran
> Through caverns measureless to man,
> Down to a sunless sea. ...

Purchas his Pilgrimage was Samuel Purchas's first book, and a best-seller. The title made Purchas a brand name, and he used it to advantage by devoting the next ten years of his life to bringing out new editions, and then compiling in five volumes the even more massive bestseller, *Purchas his Pilgrimes* (Fig. 18). His biographer declares it to be 'the largest book ever seen through the English press'. To get the materials he needed to fill his books, Purchas turned to friends and acquaintances, among them John Selden. The two men had much in common. Neither had eminent families or noble connections – Purchas's father was in the cloth trade – yet both found their way in literary London on the strength of their intelligence, determination and whatever connections they made at university (Selden at Oxford, Purchas at Cambridge). The older of the two decided to make a safe living as a vicar, the younger a less stable career in law, but neither wasted all that much effort on his career. Their enthusiasms lay elsewhere.

The two men were brought together in London by their common passion for learning and the many friends they had in common. They were close enough by 1613 for Purchas to thank Selden, 'that industrious and learned gentleman', for providing him with material for *Purchas his Pilgrimage*, and Selden could return the compliment by contributing two poems to the front of the book, plus a long note praising Purchas for using Selden's historical method of tallying accounts in the Bible against other historical sources. In personality they differed. Selden was exact, reflective, even finicky; Purchas casual, extravagant, sloppy. The differences put their friendship on a collision course.

After the 1617 edition of *Purchas his Pilgrimage* came out, Selden discovered to his dismay that Purchas had 'maimed' (his word) the essay he had contributed on the history of the Jews in England, resulting in an account less sympathetic than the one he had written. Purchas never corrected the text. The two did not break off contact entirely. They would have had to tolerate each other as members of the Virginia Company (like the EIC, another Elizabethan merchant trading monopoly enjoying state privilege) after Purchas was admitted in 1622, although Selden became inactive in the Company's business shortly thereafter. Purchas's removal of two poems by Selden from the final 1626 edition of *Purchas his Pilgrimage* signalled the end of their friendship. Purchas died that year before the breach could be healed.

Volume 3 of *Purchas his Pilgrimes* contains a great deal that would have appealed to Selden: the journal of John Saris, the reports of Richard Cocks, the exploits of Will Adams, to mention just a few. Given his intense interest in de Groot's defence of the Dutch monopoly in the Spice Islands, it would be strange had he not read these portions. Selden read everything.

It is in this volume that Purchas prints two maps of China. The first is entitled *Hondius his Map of China* (Fig. 19). It comes from the world atlas that the Amsterdam cartographic publisher Jodocus Hondius produced in 1608. Turning China 90° to the right, so that west is at the top of the map, was a design decision by the great commercial cartographer Abraham Ortelius. Purchas's copy of Hondius's map is an almost exact replica of the map Ortelius published in 1584, minus a few stray elephants and decorative sea creatures. It was re-circulated again in John Speed's world atlas of 1627, *Prospect of the Most Famous Parts of the World*. Speed turned China 90° to the left so that north was again at the top of the map, which is where subsequent cartographers left it.

Purchas includes Hondius's map for a perverse purpose: to demonstrate 'the erroneous conceits which all European Geographers have had of China'. He inserts the map at the beginning of a description of China by the Jesuit missionary Diego de Pantoja. Pantoja begins his description of China with a declaration: 'This great Kingdome of China is almost foure square, as the Chinois themselves describe the same.' The Portuguese Jesuit knows that China is square, and European cartographers don't, or so Purchas decides. Having thus exposed the error of European maps of China by singling out Hondius, he then promises that his readers will see a 'more complete Map of China' later in the volume. The correct map of China is duly unveiled forty pages on, and Purchas makes a great fuss about it (Fig. 20). This is not a European map of China, he stresses, but a Chinese map. It shows what China actually looks like. Europeans, 'knowing nothing of them, have entertayned, and beene entertayned with Fanci-maps, instead of those of China'. The time for fanciful maps was now at an end.

Purchas explains that the original of the Chinese map came to him from none other than John Saris, who acquired the map in Bantam under awkward circumstances. The owner was a Chinese businessman who had to forfeit his goods to the East India Company for debts unpaid. Saris, 'seeing him carefull to convay away a Boxe, was the more carefull to apprehend it, and therein found this Map, which another Chinese lodged at his house, lately come from China, had brought with him'. The man had tried to spirit the map away on the understanding that foreigners should not be permitted to have maps of China. (My experience at Friendship Pass thus belongs in a long tradition of embargoes on national maps.) 'The greatnesse of the danger at home (if knowne) made him earnestly begge for that which was on the other side as earnestly desired and kept', Purchas notes. Saris refused to relinquish the map, and when he left Bantam in 1609, he brought it back with him to London.

The map passed first to Richard Hakluyt, a cartographic adviser to the East India Company and Purchas's predecessor in the publishing of travellers' tales. Hakluyt supplied Purchas with much of the material that went into *Purchas his Pilgrimes*, as the latter acknowledged by giving the book the subtitle *Hakluytus Posthumus*. Purchas would have acquired the map from Hakluyt's estate after he died in 1616.

Purchas admits to being at a disadvantage in presenting this map to his readers, as he can't read the Chinese labels. 'It being in China Characters (which I thinke none in England, if any in Europe, understands) I could not wholly give it, when I give it; no man being able to receive, what he can no way conceive.' He is not dismayed but, rather, delighted, for it gives us 'nothing lesse than China in their China' and not China as Europeans have conceived it. The image that European geographers have conveyed of China is all wrong, he insists. 'They present it in forme somewhat like a Harpe, whereas it is almost foure square' – which he gets from Pantoja, as we have seen. He commends the industry of European cartographers in trying to generate a map of China, 'but industry guided by fancie, and without light, is but the blind leading the blind'. Now at last, unencumbered by the devices of 'European Art', the reader has before his eyes 'a true China, the Chinois themselves being our Guides'.

In the empty spaces at either side of the map he inserts vignettes of 'a Chinese man' and 'a China woman' with 'their little Eyes and Noses, long Hayre bound up in knots, womens feete wrapped up, long wide-sleeved Garments, Fannes, &c.'. He assures us these are true likenesses rather than wild guesses, for he has drawn them from an album of pictures 'made also in China in very good Colours' – and provided, again, by Captain John Saris (who brought home more, it seems, than erotica). The cameo of Matteo Ricci, the founder of the Jesuit mission and the first publisher of European maps in China, he got from the Jesuits.

The printed map that Saris acquired was large. Purchas notes that it measured almost 4 foot high and 5 feet wide. The original was bordered with panels furnishing practical information about each province rather than the vignettes with which European printers decorated their maps. European viewers expected instructive pictures; Chinese expected useful data. Purchas has removed the place-names 'because we exactly knew not their meaning'. As he explains, 'silence seemed better, than labour to express an unknowne Character, or boldnesse to expresse our owne folly or to occasion others, deceiving and being deceived.' However, he leaves the little boxes on the map that were the labels for the cities, so these could in theory be reconstructed. He includes only provincial names, although, being uncertain about correct spellings, he opts for one set of renderings without claiming these to be accurate, adding in his defence, 'I durst not interpret all, chusing rather to give an uncertayne truth, than to hazard a certayne errour.'

Purchas has given the map an English title: *The Map of China*. Wanting as well to give his readers 'a taste of the China Characters', he has distributed the English words within the original. Neither of the English nouns he chooses actually appears in the original Chinese title, *Huang Ming yitong fangyu beilan*, which may be translated as *The Unified Terrestrial Realm of the Ming Empire Complete at a Glance. Beilan*, 'complete at a glance', was standard advertising copy among commercial publishers, who promised non-scholarly readers that they would get all they needed in a single, affordable format that left nothing out. *Fangyu* is an ornate but still conventional way to designate not China but the earth. The earth was square (*fang*) to the four directions, and humans rode on it as they would ride a cart (*yu*). *Yitong*, 'unified under one rule', is a euphemism that the Mongols used to describe China when they conquered it in the thirteenth century – under none other than Coleridge's 'Kubla Khan'. And since the Mongols 'unified' China to make the Yuan dynasty, the Ming, not to be outdone, claimed the identical achievement for themselves. So the language of unification stuck, and indeed is still a term in common use today to designate China's national space. Finally, *huang Ming*, the simplest part of the title: unpacked, it means the imperial (*huang*) dynasty called the Ming. *The Map of China* fails to translate any of this, but then Purchas had no one to ask what the original title actually said. Better, as he put it, 'to give an uncertayne truth'.

The map from which Purchas worked has disappeared. Such wall maps were common at the time and relatively inexpensive, especially in the province of Fujian, which was the centre of mass-market publishing in the Ming. Unfortunately, no other copy of this map has survived the intervening four centuries in China. The frailty of paper is not the only reason; the change of dynasty in 1644 was also a factor. The Manchus, who invaded from Siberia to rule China under the Qing dynasty, were uneasy about the legitimacy of their conquest. They had to undermine the lingering appeal of the native dynasty they ousted if they were fully to replace the Ming. One way of doing this was to outlaw all representations of the old dynasty. Accordingly, only a fool would keep a *Unified Terrestrial Realm of the Ming Empire Complete at a Glance* up on the wall; only a greater fool would not get rid of it entirely, lest some nosy neighbour decide to denounce him to the new authorities for the capital crime of wanting to deny the legitimacy of the Qing and restore the old regime. The odds of this map surviving self-censorship at the change of dynasty are thus almost nil. The maps that have survived have done so almost entirely outside China.

————————

John Speed was London's most successful map publisher during the reign of James I. His *Prospect of the Most Famous Parts of the World* – the world atlas in which he reproduced Hondius's map of China that Samuel Purchas so disdained and which he rotated so that north was at the top – would be England's authoritative account of world geography for years to come. Speed started his working life on a lower rung of the social ladder than did Samuel Purchas: he was a Cheshire tailor, the son of a Cheshire tailor. Speed broke from his fate by moving to London in his late twenties and plying his trade there. He also indulged in what fellow members of the tailors' guild referred to as a 'very rare and ingenious capacitie in drawing and setting forthe mappes'. His first published map, of Canaan in Biblical times, a considerable production printed on four large sheets, caught the attention of the poet and parliamentarian Fulke Greville. Greville became Speed's patron, finding a sinecure for him in the customs service – conferred on him by Elizabeth I – so that he could put tailoring aside and pursue a full-time career as a cartographer and antiquarian. Speed later thanked Greville in print for 'setting this hand free from the daily imployment of a manuall trade, and giving it full liberty thus to express the inclination of my mind'. Just as importantly, his patron sponsored him for membership in the Society of Antiquaries. This organisation, founded by the Westminster schoolmaster William Camden and the virtuoso scholar Robert Cotton in 1586, was the centre of cutting-edge scholarship in the 1590s, when Speed joined. Both Camden and Cotton would take personal interest in Speed's work, Cotton in particular by giving him free access to his extraordinary manuscript library, which he would also open to Selden a decade later.

Thus it was that the son of a tailor was drawn into the ranks of the great scholars and poets of the age. With their encouragement he took on ever grander projects. His first major commission was to produce the maps for the King James Bible, which was issued in 1611. The next year he published *Theatre of the Empire of Great Britaine*, Britain's first truly national atlas.* Sixteen years later he brought out his *Prospect of*

---

* Speed outsourced some of the work for his *Theatre of the Empire of Great Britaine* to none other than Jodocus Hondius, including some shire maps, which he printed and sold individually to recoup costs during the long process of producing the atlas. The East India Company sent some of these with John Saris. Richard Cocks lists 199 of them in the inventory of the Hirado factory. Small world.

the Most Famous Parts of the World, the grandest atlas Britain had ever
seen.

That the son of a tailor – like the son of a cloth merchant (Purchas)
or the son of a fiddler (Selden) – could make this social leap was a conspicuous aspect of what made the Elizabethan and Stuart age unlike any
other in English history – and glaringly unlike its contemporary, Ming
China. Luo Hongxian, the author of the original from which the map
Purchas printed was derived, was about as far from tailoring as anyone
could be. He started out in life with all the advantages of someone whose
family was prepared to propel him up the ladder of government service.
His father had passed the highest civil service examination, winning the
title of *jinshi* (Presented Scholar) in 1499, and had gone on to enjoy a
satisfactory if unspectacular career. Luo passed the *jinshi* exam himself
in 1529 at the young age of twenty-five and was immediately seconded
into the Hanlin Academy, the Beijing think-tank that wrote policy recommendations for the emperor. The Academy was where the Ministry of
Personnel put the smartest people coming up through the examination
system, men whose talents would be wasted if they were sentenced to
careers as bureaucrats.

So far, so good. Then Luo's career was stalled by his father's illness
and death in 1533. Mourning obliged him to take a three-year sabbatical.
Before that period was over, his mother died, and another three years was
imposed. Not until 1539 was he reappointed. Within two years Luo was
dismissed for offending the emperor. He had had the temerity to suggest
that, since the emperor never appeared for the daily morning audience
with his officials, the heir apparent should be invited to take his place, at
least at the New Year's audience. The proposal may strike us as reasonable, but the emperor took it as Luo may actually have intended, as suggesting that an emperor who did not do his job should abdicate. He was
fired, and his name was removed from the civil service register. Not only
was he no longer an official; effectively he had never been one. It was a
stiff penalty for someone who anticipated a life of state service. Seventeen
years after being struck from the record, he would be recommended for
reappointment by the highest civil official in the land, the Senior Grand
Secretary, this time to the Ministry of War, but he declined.

What was a bright man blocked from advancing up the official ranks to do? The usual course was to go home and become a teacher, which he did. His teaching, however, went against the subjectivist current popular at the time. 'Within one's capacity,' he wrote, 'one should deem everything related to society or public affairs as being his responsibility, and should accept it without hesitation.' Even if you lost the franchise of public service, as Luo had, you were still obliged to be proactive in the public interest. This was not easy, since only those on the civil service register had the legal right to make policy recommendations, and then only to the local magistrate according to strict rules of precedence. Step outside that process, and you exposed yourself to the charge of concerning yourself illegally with state matters – and ultimately of questioning the decisions of the emperor.

Luo's first stab at acting in the public interest was to propose a reassessment of taxes in his home county in the province of Jiangxi, one province in from coastal Fujian. He believed that taxes fell disproportionately on the poor, and that his position of privilege obliged him to intervene on their behalf and redistribute the tax burden equitably. The suggestion required resurveying all the agricultural land in the county, a monumental task that no career-oriented magistrate would dream of wasting his time on, especially as it promised to get all the rich people up in arms. Luo volunteered to do the work. It was a gruelling task that took him six years to complete. Measuring fields may have been the practical experience that inspired Luo to take on the much bigger project of resurveying the entire country. He didn't actually go that far, but on the basis of his survey work he decided that he had the methods to redraw the map of China. This led to the achievement for which Luo is now celebrated, the publication of an atlas of forty-five regional and provincial maps, plus one national map, entitled *Guang yutu* or *Enlarged Territorial Atlas* (Fig. 25). China's first comprehensive national atlas was published in 1555, seventy years before Speed issued his.

One of the conspicuous features of Luo's atlas is his single-handed revival of the method of drawing maps by laying them across a grid of uniform squares. The technique in China dates back to at least the third century, although its full potential for accurate cartography was not fully realised until the fourteenth century, when the cartographer Zhu Siben produced a stunningly accurate gridded national map 2 metres square.

Grid-mapping fell so far out of use after the Yuan dynasty that it took Luo three years to locate a copy of Zhu's national map. It came without instructions but was enough for Luo to refine the technique and use it to produce the *Enlarged Territorial Atlas*. The atlas was a commercial success and became the industry standard for all subsequent cartography of China, including the map that Saris seized and Purchas reproduced in *Purchas his Pilgrimes*.

The Saris map lacks Luo's grid, presumably because the original lacked it. Chinese map publishers preferred to omit the grid, which some regarded as aesthetically unpleasing. For Europeans, by contrast, a grid of latitude and longitude was what they now expected on their maps: the grid was the signature of accuracy. So Purchas obliged, deciding 'to adde Degrees to help such Readers as cannot doe it better themselves'. This addition looks nice, and since the lines curve, they create the illusion that Luo drew the original according to a curved projection, which he didn't. By gilding the lily in this way, Purchas tried to lend the Saris map even more authority as the correct map of China. What tells him he is right is the secret he learned from the Jesuit missionary Pantoja, which we have already read: that China is 'almost foure square'. But is it?

———————

Zhang Huang was from the same province as Luo Hongxian. They were men of different generations. Luo was already twenty-three by the time Zhang was born in 1527, and by reaching the remarkable age of eighty-one, he would outlive Luo by forty-four years. Being of the next generation, Zhang was part of the cultural and intellectual ferment during the first half of the reign of Emperor Wanli (1573–1620). Luo Hongxian did not live to see this tumultuous era, when elite taste veered away from the intellectual and moral conservatism of Confucianism. The prize of learning still remained a post in the civil service, but the prosperity created by a burgeoning commercial economy pushed many more young men onto the bottom rung of the examination ladder than it could accommodate. Many looked elsewhere – to Buddhism, some even to the Christianity that Jesuit missionaries were trying to introduce to China – to make sense of the world in rapidly shifting times.

Zhang Huang responded creatively to these challenges in part because, unlike Luo, he failed the exams. The path to public service was thereby closed. This failure brought the benefit of insulating him from the crushing factionalism and opportunism of court politics. More importantly, it left him free to become one of the great scholars of his generation. Rather than disappearing down the rabbit hole of writing commentaries on the classics or wearing himself out schoolmastering students for the examinations, he embarked on several major enterprises. His views on the world were increasingly solicited, so in 1567 – the year the emperor allowed Moon Harbour to re-open for foreign trade – he built Purification Hall, where on the 25th of every month he lectured on topics ranging from Confucian philosophy to the workings of the cosmos. Within a year hundreds were showing up to hear the great man thinking aloud.

At least a decade before he began his lectures, Zhang had started working on a massive project to amass all knowledge of nature and society into a compendium that would summarise the state of the world. It would take him twenty years to compile. Zhang was so focused on this work, it was said, that 'he hung a lamp above his desk and did not let his pen fall from his hand summer or winter, day or night, for many years running. A swarm of mosquitoes could descend on his limbs and body and he wouldn't even notice them.' The title, *Tushu bian*, means 'Compendium of Texts and Illustrations', but I call it the *Documentarium*: a virtual aquarium of essays, documents, illustrations – and maps.

The *Documentarium* came to occupy roughly the same position in the world of Ming books as *Purchas his Pilgrimes* did for the English reading public. The books' content is different, but both authors aimed their material at a broad readership. And both clearly succeeded, for their books were widely read for generations. Both were the work of men in their forties, although the two men met different fates. Purchas outlived the publication of *Purchas his Pilgrimes* by barely a year, dying just shy of fifty and just short of bankruptcy. Zhang was exactly that age when he completed the draft of the *Documentarium*, but he lived for another thirty-one years, eventually becoming head of the White Deer Grotto, the most prestigious private academy in Jiangxi Province, indeed in the country, in 1590. Yet even though he lived to the great age of eighty-one, he did not have the satisfaction of seeing the *Documentarium* into print, dying five years short of its publication date in 1613.

Zhang Huang was not the commercial operator that Samuel Purchas was; nor was he the social activist that Luo Hongxian had been. But he outmatched both as a scholar. Luo spotted a problem and burrowed into it as deeply as he could. The *Enlarged Territorial Atlas* was the outcome of just such concentration. He was the kind of scholar who insisted on knowing the practical effect of his research. In the case of his atlas, it was to improve geographical knowledge in order to increase security. His own county had once nearly been overrun by bandits, and he believed that a better grasp of local topography would have contributed greatly to knowing how to deal with the problem. Geographical ignorance impeded effective administration.

Zhang took a less goal-oriented view. The duty of the scholar was to amass the best knowledge and to make it available to those faced with solving real-world problems. As he put it in another book he published at sixty, the obligation of the researcher was not to indulge his own views. It was to achieve reliability. 'What I have striven for is to make present knowledge reliable and then pass it down to later ages.' As new knowledge accumulated, older knowledge had to be revised and thinned out accordingly. New facts 'may then be added, but only when there is documentation adequate to verify them. I have not dared to rely on my own views whatsoever to discriminate one fact from another.' Rather than dare to 'give an uncertayne truth', as Purchas was sometimes willing to do, Zhang strove to hold to the strictest standards of objectivity.

Physical and historical geography take up almost a third of the *Documentarium*, a percentage that has guaranteed the inclusion of dozens of maps. One finds the standard representations of what Chinese thought China looked like, but there is much else as well. Zhang's *General Map of Chinese and Barbarians within the Four Seas* is particularly striking (Fig. 23). 'Maps of Chinese and Barbarians' (*huayi tu*) was a distinctive Chinese cartographic genre showing China as the heartland of civilisation, with peoples who have not had the benefit of civilisation jammed in unrecognisably around its edges. Zhang uses the title to invoke what is essentially the Chinese tradition of world mapping, but he goes beyond the genre by allowing China to blend into a far larger Eurasian continent surrounded by oceans on all four sides. This may not be Eurasia as we know it, but Zhang imagines the continent more coherently and forcefully than any Chinese geographer had previously done – and without referring to European maps. He is a little self-conscious about the map, inserting a label in the upper right-hand corner that explains to the reader that 'it has been included in order to facilitate further research'. Because Zhang is reaching way beyond what his sources allow him to know, he makes some interesting leaps. Take the huge lake in the middle of the continent named Hanhai, the Boundless Sea, which the engraver has hatched with wave marks to indicate that this is a body of water. The term was actually coined a millennium earlier as a metaphorical name for that great sea of shifting sand, the Gobi Desert, but by the Ming the metaphor had been lost. Not realising this, Zhang moves from 'uncertayne truth' to 'certayne errour' by converting the desert into a vast, non-existent lake. Despite what he gets wrong, Zhang's *General Map of Chinese and Barbarians within the Four Seas* steps forward from the cartography of his time by bringing into view Eurasia, a continent that dwarfs China and extends far beyond its borders in a way that no *huayi tu* had ever done before.

Zhang might not have included the map had he not encountered the Jesuit missionary Matteo Ricci. They met briefly, and the Italian scholar had a great impact on Zhang, who included in his encyclopaedia a lot of material Ricci gave him: maps of the eastern and western hemispheres copied from the atlas of Girolamo Ruscelli (the Latin labels have collapsed into gobbledygook); two azimuthal projections of the northern and southern hemispheres marked out in 360° of longitude; and a map of the world based on Ortelius (Fig. 21) – which Ricci later published as a giant wall map in twelve sheets in Beijing. The inclusion of these maps challenged Chinese who were sceptical about European claims about how far they travelled to get to China. Zhang was living up to his principle of not sticking to his own preconceptions when something new came along.

The *Documentarium* absorbed some of the newest cartographical knowledge from Europe, but the borrowing went the other way as well. Europeans such as John Saris who travelled to Asia and saw Chinese maps had no way to visualise China other than as these maps showed it. Recall Samuel Purchas's approving citation of the Jesuit who stated categorically that China is 'almost foure square, as the Chinois themselves describe the same'. This was indeed what people of the Ming would say about the shape of their country if asked. It was an *idée fixe* that China was a rectangle. Zhang captures that idea with the illustration that opens his first chapter on cosmography. It is a simple half-folio diagram of an elongated rectangle inside a circle, with a caption quoting the ancient adage, 'Heaven is round, earth is square'.

This was the design principle that no Chinese cartographer, not even Zhang Huang, could entirely escape. Look again at the Saris map in Purchas, which was derived from Luo Hongxian's 1555 map of China (Figs. 20, 22). The south-east coast does curve, and the Gulf of Bohai takes a bite out of the north-east coast, but the overall shape is roughly squared off: certainly enough to stay within the cosmic formula of a round heaven and a square earth. The fact that the word for 'square', *fang*, is also the technical name of the squares in Luo's grid only helps. All a cartographer had to do was to keep adding squares equally in all four directions until he had a complete map of the country. A large square emerges as the sum of its small squares: the place and the map produce each other.

Now look again. It is just as easy to reverse the logic and see Luo's China as not square but ovoid, which Luo squared up as well as he could. To see China as square is to see it as Chinese culture has taught Chinese to believe it to be. To see China as a different shape is to see it without that preparation – and this is what the Europeans who first saw Chinese maps in the sixteenth century did. They experienced China from its coastal edge, which is unambiguously curved. Looking at Chinese maps, which they had to, having at first none of their own, they shaped their image of China into an oval. Purchas described their ovoid rendering of China as being in the form of a harp, and believed they were mistaken. But this was a mutually constructed mistake: Chinese cartographers pushed China into as square a shape as they could manage, then European cartographers rounded the lower coast to conform to their maritime experience, following which mapmakers back in Europe then incorporated this massaged version of the Chinese map of China into their world maps. These operations produced the image of China in Ortelius, then in Hondius, then in Speed, and then in most of the great atlases of the seventeenth century. Purchas's instinct to trust the unmassaged Chinese version on the Saris map is a good one, just the sort of instinct that always kicked in with antiquarians such as Purchas and Speed and Selden: always go to the source closest to the time and place of whatever it is you are investigating. It's just that Purchas did not have available to him all the steps in the logic that distorted both Chinese and European maps in this period. There was no true or objective image of China at the time, only the images that culture and experience shaped, and not always in ways that everyone spotted.

At the heart of the matter is the geometrical challenge that drove the development of modern cartography: the problem of how to relate the curved to the flat. For Zhang Huang this wasn't a problem he had to solve. He saw the relationship between circle and square as a cosmic pattern that shaped the disposition of the land he mapped. For European cartographers in the sixteenth century, however, the problem was the curvature not of heaven but of earth. And it was a problem they had to solve because their navigators needed it solved.

The simplest solution to the problem of the earth's spherical shape was to draw the world on a globe. Will Adams did this for the shogun of Japan, who at first thought Adams was lying to him but in the end was mightily impressed. Adams wrote to the East India Company asking that they send him a pair of globes in order to persuade the shogun to support England's bid to open a North-East Passage over Russia that would shorten the route between them. However visually impressive they could be, globes were utterly impractical for navigation. Mariners needed larger scale and finer detail than any globe could provide. They also needed something that could be stored flat.

The solution is called a projection: that is, a method for casting an image of the earth's curved surface onto a flat piece of paper in a way that respected curvature while limiting the distortion as much as possible. But distortion is unavoidable. The short answer to the problem of reducing the curved to the flat is that you can't. You can only approximately square a circle. *Pi* is the name we give to the relationship between a circle and a square. Calculating pi produces a number that never ends. Cast the problem in three dimensions and completing the calculation only gets harder. A globe didn't completely solve this problem, since the gores or strips of paper that a globe-maker pasted onto the surface of a ball had to be printed first in a flat format. Every projection is a compromise.

Mariners encountered the problem of curvature in the course of doing the simplest thing: sailing in a straight line. Draw a straight line from a fixed point on a globe and then draw one from the same point on a flat map, and you will discover that they do not end up in the same place. Map a route on the land and this is not such a problem, as the wayfinder can adjust his line constantly in relation to visual reference points. This discrepancy can be fudged over open water and the pilot has to work from what are known as dead ('deduced') reckonings, which he deduces on the basis of compass angle, wind direction, current, duration of travel and where he was at his last dead reckoning. Miscalculate one dead reckoning and the next must be at least as inaccurate, which only increases the overall degree of error.

Sailors in the Mediterranean tried to achieve geometrical control over maritime space by using portolan charts. By deploying a dense web of magnetically fixed lines (rhumb lines) radiating from a grand circle of compass roses, the pilot had a potential infinity of lines against which to set his course on paper. But curvature defeated the assumption that magnetic bearings produced consistency. North, south, east and west never varied, but every other position on the compass produced tangents that went off direction as the earth curved. This didn't pose problems over short distances, especially when the ready availability of coastal bearings allowed for tiny corrections all along the way. What killed the portolan tradition was long-distance voyaging over blue water, when mariners discovered that sailing on a fixed magnetic bearing didn't take them where they thought they were heading.

This is where Gerard Mercator enters the story of European cartography. Born Gerard Kremer, he latinised his surname when he started publishing his work. (Kremer is the Dutch equivalent of 'merchant' or, in Latin, mercator.) He began his working life as an instrument-maker rather than a tailor, yet he anticipated John Speed as a mapmaker by starting out with a six-sheet map of Palestine, the Holy Land of the Bible. It was a predictable point of origin, understanding the truth of the Bible being the first task of the antiquarian. He quickly grasped that the secret of producing an accurate flat map for a spherical earth was to work from the sea and not from the land. The sea was where the problem of constant direction was more acute; it also provided the cartographic innovator with a large unmarked space in which to work out a solution to what was a purely mathematical problem. Mercator started to develop that solution in 1541 by drawing rhumb lines on a terrestrial globe and considering the consistencies in how those lines curved – for it is in the nature of a sphere that every line drawn on a constant compass bearing turns out to be a spiral, ending at either the North or the South Pole.

The challenge for Mercator was how to draw this spiralling line so that it looked as straight on a flat map as it felt to a mariner when he was keeping on a constant compass bearing. His solution was ingenious. Rather than bend the line to accommodate the actual shape of the land, he chose to distort the land. This distortion required stretching the globe towards each pole, increasing the degree of stretch at the rate of pi as he approached each pole. This north–south elongation entailed a proportionate east–west stretch that also increased the further one moved from the equator. Bend the model around on itself, and the earth became in effect a cylinder. Mercator got to his model through empirical experimentation but was sufficiently well trained in mathematics to detect its geometric regularity. The distortion he needed in order to ensure that sailing routes appeared as straight lines was not random but could be worked out by precise calculation. The result is what we now call the Mercator projection.

The appeal of the Mercator projection was that it worked for navigators, and in the sixteenth century they were the ones sailing across vast spaces and needing reliable maps. Mercator simplified the challenge of plotting a course by drawing the world in such a way that any point A and point B could be connected by a straight line that stayed on a single compass bearing. The actual course the ship followed was in reality curved and therefore not the shortest distance between two points, but it could be calculated simply and reliably by by locking a ship's bearing on a fixed compass direction. The disadvantage of taking a slightly longer route was offset by the certainty that you would arrive at your destination. Mercator had squared Zhang Huang's circle.

Mercator tried to put his projection into popular circulation with his massive world map of 1569. It didn't take immediately, but by the end of the century it had become standard. Even today Mercator's redrawing of the world is the image most of us carry around in our heads: Canada,

Greenland and Russia inflated beyond their actual sizes, and Antarctica distorted into a titanic continent, like Atlas's shoulders, on which the world rests. Other cartographers have tinkered with Mercator's model, trying to adjust it to reduce distortion without sacrificing accuracy. Mercator's colleague Abraham Ortelius came up with a method that curved Mercator's lines of longitude on either side of the prime meridian (the central line of longitude on a map), increasing the degree of curve the further he went. The result, called the pseudo-cylindrical projection, was a compromise between square and round that reduced the distortion caused by the Mercator projection. This was the projection that Matteo Ricci used for the *mappa mundi* he drew for Chinese viewers, and which Zhang Huang reprinted in his *Documentarium* (Fig. 24). Copies of copies of copies: this is how cartographic knowledge spreads.

For his regional maps, including that of China, Ortelius tried yet other projections. His data were others' renderings of China and therefore too imprecise to permit the construction of any sort of precise projection. He does insert a latitude scale across the top and bottom of the map, inferred from other sources, but that is as far as he can go. Via Hondius, this becomes Purchas's 'bad' map of China (Fig. 19). To prop up his 'good' map, the Saris map, Purchas applied a grid of latitude and longitude over it based on Portuguese data (Fig. 20). In so doing, however, he discovered that the longitude values in the sources available to him did not agree with 'the generall opinion'. Purchas suspected that the misfit had to do with the competition between Spain and Portugal in the wake of the Treaty of Tordesillas in 1494, which claimed to divide the globe between the two states. Because the division was based on longitude, it mattered deeply what longitude readings were ascribed to places over which these states wanted control, even before they actually got there. But the Iberians aren't entirely to blame. Earth being square, Chinese cartographers had already imposed a template that stretched China out of its natural shape, dragging Beijing, for example, further east than it should have been. Purchas's fate was to mash together two incompatible geometries in order to come up with what he thought was a true map of China. The unfortunate result of this laudable attempt was that the distortions of one made worse the distortions of the other. Purchas's rendering of the Saris map ends up being his own heroic creation, a hybrid that looks as suspiciously accurate to Europeans as it does to Chinese and yet falls short of both standards.

We can't blame Purchas. He created his mindscape of China on the basis of what he believed. It was simply an image, after all. No navigator would have dreamed of sailing to Coleridge's sunless sea with only the Saris map as his guide, for the simple reason that no navigator had sailed to China to draw it.

――――――――

Oscar Wilde once famously quipped that 'a map of the world that does not include Utopia is not worth even glancing at'. If Xanadu was Coleridge's Utopia, he wasn't going to find it on either of the maps of China in *Purchas his Pilgrimes*. Even if Purchas hadn't emptied all the little rectangular labels festooning the Saris map, Xanadu wouldn't have been there, for by the fourteenth century it was a ruin that the Ming had abandoned to the Gobi Desert. Not so the Selden map. There it is, in the north-east corner of China, mysteriously inscribed inside a gourd instead of a circle: 'Upper Capital of the Jin and Yuan dynasties'. 'Upper Capital' in Chinese is Shangdu, which in Purchas became Xamdu and was filtered through Coleridge's fertile imagination – and his need to stretch a spondee (two stressed syllables, XAM-DU) onto the first half of an iambic tetrameter (in-XA na-DU did-KU bla-KHAN) – into Xanadu. The name is right, as are the historical references the map supplies. The Jurchens established the Jin dynasty and built a capital here at the southern edge of the Gobi Desert in the twelfth century. Khubilai Khan rebuilt it in 1256 as his capital, only to abandon it nine years later in favour of Beijing, six years before formally declaring the founding of the Yuan dynasty. What isn't right is the location. Xanadu lay 300 kilometres directly north of Beijing. On the Selden map it lies about twice that distance to the east. Once again the map is 'wrong', but that doesn't give us the right to chastise the cartographer. Xanadu disappeared three centuries before he drew the map. This may explain the gourd-shaped label. Gourds were the Aladdin's lamps of Chinese imagination, out of which could emerge fantastical visions. By using a gourd, the cartographer has marked it as a place in the imagination, a place that did not exist – which is literally what the English 'utopia' means.

Wilde would have approved, and so should we, for Xanadu will prove to be one of the trace elements that allow us in the final chapter to reconstruct just how the Selden map got drawn in the unusual way it did.

8
Secrets of the Selden Map

Ever inventive, endlessly ambitious and always short of cash, Ben Jonson rose to the pinnacle of public fame by abandoning serious writing for the seduction of spectacle. James I and his court paid handsomely for his masques, and in 1620 he dutifully churned out yet another Cirque-du-Soleil extravaganza. The plot of *News from the New World Discovered in the Moon* is meagre, little more than an excuse to amuse his self-regarding king with a lot of dancing and singing. Still, there had to be a story to pique the court's interest, so Jonson has two heralds come out on stage early in the show to announce that – surprise! – moon people are about to arrive in England.

Displaying strange people from abroad gave

Jonson a choice. He could fashion them into figures exotically strange, even barbarous, for which the Americas supplied examples for his art. Or he could make them like his audience, civilised, though perhaps touched with the mildest difference. Chinese, for example: they were civilised. The dancers in the masque being from the moon, Jonson could do anything he wanted. He chose to make them civilised. The heralds proclaim that the moon has been 'new found to be an earth inhabited, with navigable seas and rivers, variety of nations, policies, laws'; in other words, just like Europe – 'but differing from ours', they add. Difference is important, or rather its degree. Entirely the same, and the company of moon dancers about to appear would have nothing exotic about them. A little difference was essential to create a bit of excitement on stage; at least their costumes should be outlandish, perhaps their language gibberish. Utterly different, though, and the moon people would have to act out the threat the audience perceived in them, producing a very different theatrical mood: a battle rather than a comedy.

Eight years earlier, in 1612, William Shakespeare handled the dynamic of intercultural encounter differently when he staged *The Tempest* at court. Instead of civilised moon people coming to civilised England, civilised Europeans arrive on a savage island. 'Uninhabitable, and almost inaccessible' is how one of the characters describes what he sees around him. This is a land utterly without the 'nations, policies, laws' that Jonson imputed to the moon. The island's sole native inhabitant is Caliban, 'a freckled whelp hag-born – not honour'd with a human shape'. Caliban is enslaved by the central character, Prospero, who was driven from Milan by his usurping brother and shipwrecked on this island. Shakespeare gives his audience a violent Caliban worth loathing, and yet he lets Caliban tell his story: that he started as an innocent, a natural man who welcomed Prospero at the moment of first contact. It was Prospero's expropriation of the island that turned Caliban into what he became that created the monstrosity of colonialism. As Caliban reminds his violent master,

You taught me language; and my profit on't
Is, I know how to curse. The red plague rid you
For learning me your language!

Even the plague Caliban calls down on Prospero was an effect of the Europeans' invasion of this *terra nullius*, disease turning Paradise foul. In the next act Shakespeare sketches the two scenarios that could follow from a *terra nullius* claim. The idealised vision he gives to the old counsellor Gonzalo, who imagines the perfect 'commonwealth' he would create on this empty island: a regime without commerce or state, contract or private ownership, labour or war. He is mercilessly mocked as an old fool by the other shipwrecked courtiers, who turn his vision inside out and declare his new golden age to be nothing but a society of 'whores and knaves' over which Gonzalo would simply end up declaring himself king. Theirs is the other interpretation: that people placed in a state of nature will pursue their own selfish interests, which of course – and the audience picks this up – is what these courtiers have been doing ever since they deposed Prospero in pursuit of their own gain. We can admire Gonzalo's sincerity, but we are under no illusion that *terra nullius* can ever work out well, even when the despot is benevolent. The play ends ambiguously with the triumph of civilisation over savagery as Prospero's lost daughter Miranda, finding herself at last among her own civilised kind, exclaims,

> O, wonder!
> How many goodly creatures are there here!
> How beauteous mankind is! O brave new world,
> That has such people in't!

The Tempest is one of Shakespeare's great plays, whereas *News from the New World* will never rate highly in the oeuvre of Ben Jonson, but then Jonson wasn't trying to write serious theatre. I offer both here at the start of the final chapter of this book because they present us with the extremes between which Europeans managed their encounters with the larger world, whether from the deck of a ship, as John Saris did (he was sailing to Bantam when *The Tempest* was being performed), or from the pages of a manuscript, as John Selden did (he was preparing to pass the Bar examination that year). *The Tempest* understands the brave new world to be a place of profound difference, a savage land demanding submission or exile. *News from the New World* seeks familiarity in foreign places and accepts that the laws and customs of other places can vary from ours without threatening us. Saris and Selden took the Jonsonian approach. The nations and peoples of the world differed, but not in essentials. Saris could go to them to trade without conquest, Selden to delve into their documents in search of the common wellsprings of enlightened humanity. It would be another century before this sense of equality gave way to condescension and the East India Company concentrated its efforts on stripping the world of its assets and other peoples of their dignity.

Jonson and Shakespeare, either way, gave English audiences what they had a taste for: visions of the new worlds that lay far from England's shores. In so doing they were simply playing on the fad for travellers' tales, the genre that Richard Hakluyt first exploited in the 1590s, that Samuel Purchas retailed through the 1610s and 1620s, and for which John Speed provided maps and illustrations. It was great entertainment and would continue to be so all the way down to Coleridge dozing in his chair at the end of the eighteenth century. But news from new-found lands had potent effects in other registers as well. Scholars such as John Selden greedily scooped up whatever they could learn about other places and traditions as that information arrived, mastering the languages and collecting the manuscripts they would need to excavate the deeper realities of human history that lay obscured from sight.

As this new knowledge arrived, nothing changed, at first. The world just got fuller. But gradually, as evidence of other ways of being and thinking came more insistently into view, some realised that the old ways were not the only ways, and indeed might have to be revised or superseded. To be alive in John Selden's day was to live through this shift in paradigms. Some – Selden, conspicuously – moved with the tide and employed the new insights to lay a stronger comparative foundation under European knowledge. Others were idled by the changes, uncertain of how to respond to what the world was showing them. Yet others were entirely left behind, stuck in old assumptions that they remained content to accept well after these unexamined ideas had collapsed under the weight of their untenability.

Even the best and the brightest could find themselves caught halfway between reaffirming the old and digesting the new. James Ussher,

Selden's teacher of Hebrew and Arabic and later his friend in scholarship, turned to ancient Hebrew texts in order to date the creation of the world to the early hours of 23 October 4004 BC. He embraced the new sources and experimented with the comparative method that the greater knowledge of the world called into being, yet he drew from them merely to confirm the Biblical account rather than to challenge it. It wouldn't take long before scholars in the next generation found abundant evidence in Oriental texts, especially Chinese texts, showing that the chronology of human history extended back well before 4004 BC. Had Ussher been paying closer attention to his sources and less to his own assumptions, he could have saved himself from this exercise in pious futility – although in that case no one today would have heard of him. The Biblical account of the creation of the world was only one casualty of the global enlargement of knowledge that inspired some thinkers to pry up the theological floorboards of European thought. This is why the smart people were learning Hebrew, Arabic and, eventually, even Chinese. There was important information buried in those ancient languages; the codes had to be cracked. Orientalists were the hackers of their generation.

Selden may have appreciated his 'Mapp of China' as evidence of advanced knowledge on the far side of the world. But did he regard it as a document that needed to be decoded? Was there something in it that he absolutely had to understand? As he never commented on the map, we have no way of knowing. Still, I am tempted to speculate that he did, that within this image of the far end of Asia he sensed there were things worth learning that he could not have otherwise deduced.

The best I can offer in defence of this claim is a passage in *Titles of Honor*, Selden's highly regarded study of the history of aristocratic ranks and privileges, and his first major scholarly book before his *Historie of Tithes* got him into trouble. The passage is not in the original 1614 edition but in a dedication he added to the second edition, of 1631. There he observes that 'all Isles and Continents (which are indeed but greater Isles) are so seated, that there is none, but that, from some shore of it, another may be discovered.' Ready communication is to be expected among islands: it is a simple observation. But is it an idea that European experience naturally brings to mind? Contemplate a map of continental Europe and there is little to inspire this insight. Yes, there are a few

(British) isles at the edge of Europe, but these have a history of resisting those coming from other shores. Contemplate the Selden map instead and you find yourself gazing at a mosaic of islands, peninsulas and lesser continents that are indeed 'so seated' that none can be isolated from any other.

From this observation Selden proceeds to consider two arguments that could be made to follow from it. The first is that 'Some take this as an Invitation of Nature to the peopling of one soil from another.' This effectively restates the *terra nullius* argument about a legal right to occupy foreign territory that is found to be vacant. The second, and more aggressive, corollary is that 'Others note it as if the Publique Right of Mutual Commerce were designed by it.' Here we come face to face with de Groot's natural-law view that commercial exchange is natural and therefore lawful, and that anyone who impedes that exchange may lawfully be challenged. We know that Selden was sceptical of both claims. In his view, the rights to occupy territory and to engage in commerce could be invoked only when other conditions were met. One of those conditions was parity: one party could not impose terms of trade or unequal contracts that overrode more fundamental rights.

Selden does not insert this observation into *Titles of Honor* in order to score points against de Groot. In fact, he uses this image of the mobility of relations among islands purely as a metaphor to make another point entirely, which is to regret that the many fields of 'good arts and learning' have become severed from each other, whereas 'every one hath so much relation to some other, that it hath not only use often of the aid of what is next it, but, through that, also of what is out of ken to it'. To phrase this in our way of speaking, all disciplines of knowledge draw on and have a bearing on all other disciplines, and should not be insulated one from the other. It's all a metaphor, and yet as a reader of the map, I couldn't help but be struck by the similarity between the image he chose and the map he probably owned by then. It is hard to imagine Selden gazing at a map of Europe and coming up with this way of capturing the necessity of interdisciplinary learning. It is very easy, however, to imagine him doing just this with his map of East Asia.

Admittedly I speculate. I do so knowing that Selden could not have been indifferent to a map that meant enough to him to need specifying in his will. What it meant to him, alas, he never declared. That needn't consign us to silence, however. If he won't reveal to us how he read the map, then we will go in search of it on our own.

When I first encountered this map, it struck me as a puzzle, as it still does. The more pieces I fit into place, the more puzzling it becomes. This has not dismayed me. All maps are puzzles, coded according to the conventions of their time and the whims of their creators. To read a historical map means having to learn its codes – and ignore some of our own, lest we be tricked into mistaking the mapmaker's devices for our own. The worst thing we can do is to condescend. To a greater or lesser degree, maps are always adequate to what they were meant to do. A cartographer draws something in a particular way because that is what he intends. If he had wanted to draw it another way, he would have done so. The problem is ours, not his. When a historical map looks 'wrong' to us, it is simply because we haven't figured out its code. In fact, the 'wrong' bits can be the best places to start looking for clues about how to crack the code. Where its code and our own fail to align is where we need to look most carefully.

But I must warn you: we won't break all the codes. The Selden map is full of secrets. Only some of them will we unlock. I count six.

The first secret is China. China sticks out on the Selden map in several ways. It contains more geographical detail than any other part of the map. It has more place-names than any other zone. And visually it has its own distinctive design, a coherence that seems complete unto itself and doesn't require the rest of the map to give it sense. It fills a lot of the map without quite belonging to it.

Consider the most pronounced feature of the China section, the serpentine channels that writhe across the landscape and bind Ming China into a unified whole. At first glance they must be rivers. But no. Already we have mis-stepped, using our code for a map to which it does not apply. What look like river channels are not, in fact, rivers at all. With one exception, they are provincial boundaries. This feature is not easy to detect, as these boundaries are coloured the same as the ocean. Worse, where they reach the coast, they seem to open onto the ocean the way that rivers do. The exception is the upper section of the Yellow River beyond the Great Wall, running up the left-hand side of the map. Its source – and labelled as such (Hyde has latinised it as *huang fluvii aqua incipium*) – is a peanut-shaped lake that runs off the edge of the map, almost opening into the Indian Ocean. Once the river gets to the Great Wall, it disappears and provincial boundaries take over. The same visual system is thus used to depict two completely different things: rivers and boundaries. It's confusing, and not a common device on Chinese maps.

Also visually prominent are dozens of cities, each in a label edged in yellow, or crenellated in red in the case of a provincial capital. There are also dozens of single-character labels edged in red. At first glance they must be lesser places. Wrong again. The names in these circles are not city-names. They are concrete nouns: Basket, Net, House, Rooftop, Wall, Well, Horn, Wing, Tail, Ghost, Star, plus a few more besides. Chinese familiar with the night sky will recognise this vocabulary list, for these are the names of some of the lunar mansions into which the night sky is divided. There are twenty-eight of them, designated in relation to the movement of the moon around the earth. So what are they doing on this map, when their proper locations are up in the sky, not down on the earth? They are there on the strength of the Chinese cosmological understanding that a common field of energy, or *qi*, binds heaven and earth together. According to this theory, every place in heaven has a corresponding location on earth, or at least in China. These have been organised into a system of astral correspondences known as Field Division (*fenye*). The system dates back two millennia before this map was drawn, although by the Ming no one really understood how the whole thing worked. People knew the correspondence was there; they just couldn't explain it.

Not knowing what to do with this feature of the Selden map, I did as I usually do: I turned to my standard Ming reference work, Zhang Huang's *Documentarium*. As I flipped through the encyclopaedia looking for help, I discovered that the section before geography was about cosmology and included an entire chapter devoted to Field Division. Even Zhang had difficulty making sense of the system, so he composed his own treatise on the subject here. 'Study of the Order of Corresponding Constellations among which are Divided the Prefectures and Counties of the Realm' takes the reader through both general principles and antiquarian sources, showing how *qi* works to keep heaven and earth in sync. John Selden would have appreciated the exercise. It is just the sort of historical exploration of sources and interpretations that he enjoyed. Ultimately, though, Zhang fails. While he is able to show how the lunar mansions operate like a universal clock whose hand at noon is the angle of the sun at the winter solstice, he can't make sense of Field Division.

But Zhang Huang did not disappoint me after all. After reading his essay, I stumbled upon what I really needed: a map of China displaying regional correspondences to the twenty-eight lunar mansions. This discovery sent me rooting around in other popular encyclopedias. I found Zhang's prototype in the most widely reprinted household almanac of the era, *The Complete Source for a Myriad Practical Uses* (*Wanyong Zhengzong*), produced by Fujian's most indefatigable commercial publisher, Yu Xiangdou, in 1599. Yu's map is entitled *A General Topographical Map by Province of the Divisions and Correspondences of the Twenty-Eight Lunar Mansions of the Ming Dynasty*. It is not a perfect match with the China portion of the Selden map, but it displays most of the astral correspondences and almost the exact same set of place-names, even captions. It also turns provincial boundaries into river-like shapes, and it merges Hainan Island into the south coast.

Now for the telling detail, Xanadu. Just as the Selden map marks the old Jurchen/Mongol capital, there it is on the Yu and Zhang maps. They give the label the shape of a water drop; but like the gourd on the Selden map, this shape is reserved for this one place-name. This may seem like a minor detail, but it shows these maps to be pebbles in the same cartographic stream. This is why Coleridge deserved a nod in the previous chapter. Selden led me to his friend Purchas, Purchas led me to his latter-day reader Coleridge, Coleridge led me to Xanadu, and Xanadu led me to Zhang Huang and Yu Xiangdou. It was an odd sequence, but without Coleridge's dream-poem I might never have noticed what was going on.

The important difference between the Yu map and the Selden map is not in their details but in their frames. Yu employs the standard square format: a few place-name labels float in the ocean, but otherwise China fills the frame. Non-China barely appears. In the Selden map China is a component of a larger zone. It is also so visually distinct from the rest of the visual field in which it sits that, to me, it looks as though our cartographer has taken something like the Fujian almanac map, stripped off its frame and slotted it into the larger map like a sort of prefabricated unit to fill in the space where China is required. In other words, China does not anchor this map, which is what we might expect China to do when a Chinese mapmaker holds the pen. Instead, our cartographer has filled the space with a map he copied from elsewhere, without any serious attempt to integrate the rest of his map into it. Inserting one map into another suggests to me that he wasn't particularly interested in China. This is because the coasts mattered, not the interior.

I wouldn't go so far as to say that the bog-standard version of China on the Selden map is nothing more than decorative filler. Perhaps the Selden cartographer chose it out of an interest in the stars, given the importance of stars for night navigation. The sun's position in the daytime sky was a very blunt instrument for navigating the open ocean compared with the infinity of astral positions that moved through the night sky. Unfortunately, there is little in the documents of Chinese navigation from which to reconstruct that interest. The section on reading the stars in the Laud rutter is extremely brief, consisting only of the eight compass points at which four named constellations rise and set. Surely a pilot would command a richer repertoire of astral readings than that. Perhaps it was information that was transmitted only by word of mouth, embargoed knowledge that no pilot would want to commit to public record. We could also speculate that Chinese compass knowledge was so good that it obviated the need for detailed astral knowledge, though that I doubt. Every pilot accumulates just as much knowledge as he can to assure safe passage. How the night sky mattered I cannot say, but perhaps it prompted our cartographer to choose the map he did.*

Constellations aren't the only heavenly bodies marked on the Selden map. The sun and the moon are also depicted, twice. One pair is directly

---

* There is indirect confirmation of the importance of the night sky to navigation in the great handbook of technical knowledge of 1637, Song Yingxing's *The Works of Heaven and the Inception of Things* (*Tiangong kaiwu*). The book includes an illustration of a tax barge that depicts several simple constellations in the sky above it. The intention may have been to show that these barges travelled day and night, but it may also imply the use of the night sky for navigation.

north of Beijing: a red sun and a white moon each labelled with its Chinese character. They are repeated more prominently in the two top corners. A red sun sits in the right and a white moon in the left, both draped by auspicious wisps of multicoloured cloud. We can tell that Michael Shen pointed them out to Thomas Hyde, for the latter has written on them their Latin names, *sol* and *luna*. Pairing the sun and moon was a popular Ming device that came to an abrupt finish at the end of the dynasty. They show up as shoulder patches on the robes of the Ming emperors in the fifteenth century, for example, and they were used as a decorative insignia on tombstones. Sun and moon were understood to be the most powerful of the celestial bodies, so drawing them on the map was a device for invoking the protective power of the cosmos over the routes that sailors travelled.

There may be a sort of pun involved here too. When you combine the characters for sun on the right and moon on the left into a single character, you get the word *ming*, 'light' – the name of the dynasty. Was the Selden cartographer acknowledging the dynasty – not the place called China but the time called the Ming? If so, then had the map remained in Asia, it would not have survived the Manchu conquest in 1644. At that moment every sign of the Ming had to be expunged. The patches were torn from the shoulders of imperial robes, the talismans gouged from tombstones, and all other images of the sun and moon made to disappear as though they had never been there.

So the first secret is just that China isn't the way it appears, and that the map isn't really about China anyway. Not much, but it's a start.

———

The second secret of the Selden map is what will explain its stunning accuracy. Lay the map beside a contemporary rendering such as John Speed's *Asia with the Islands Adjoining Described* engraved in 1626 and it comes off well (Fig. 22). Put it beside a modern conical projection (the Asia North Lambert Conformal Conic Projection, to be technical) – turn to the second appendix for this exercise – and it still holds up. Map historian Cordell Yee observed two decades ago that 'scale mapping was not the primary concern of Chinese mapmakers, although they certainly understood its principles'. He made this comment to try and account for why Chinese cartography, despite its technical capacities, seemed indifferent to technical accuracy. Now that we have the Selden map, no such apology is necessary. Here is scale mapping that is the equal of the best work being done in Europe at the time. Between Speed and Selden, one gets some parts better than the other, but over all it is an even draw. What explains this?

The answer is simple, but getting there will be a bit complicated. Start with the technical problem attendant on the task of drawing a map of a large region. This can't be done without prior work. To compress vast distances into a small space you have to have an image already of what the larger region should look like. It is much easier to draw the outline of a bay than the outline of an ocean when the most you can see is the bay. Before you draw an entire ocean, you need to know what it looks like, and that you get not from your own experience but from a map you have already seen.

If the visual accuracy of the Selden map had to come from somewhere, where was that? We have already speculated that the compass rose and ruler point to exposure to a European map. It doesn't take much casting about to realise that one way the Selden cartographer was able to produce such a coherent and accurate map of the region was by copying a European map. The prototype could not have been Speed's, which post-dated it, but isn't it possible that he had seen a slightly earlier European map of East Asia and copied that? For a long time I resisted coming to the conclusion that the Selden cartographer had seen a European map before he drew his own. His map struck me, and still strikes me, as a wonderfully unique and brilliant solution to the challenge of mapping the maritime zones around China. I didn't want it to be just a copy of a European map, and was able to hold out because I could never find the map he might have copied. But I couldn't hold out for ever. Chinese drew East Asia one way, Europeans drew it another, and here was a Chinese drawing it the European way. I could no longer deny the possibility that the spatial organisation of the Selden map may have been suggested by his having seen a European map.

But how much does that exposure actually matter when the Selden cartographer gives no sign of having learned the technical methods by which Europeans measured and mapped the surface of the earth? The particular strength of European maps was their imposition of a uniform scale through the use of latitude and longitude. Our cartographer did not have access to this matrix. The variability of scale on the Selden map makes this quickly apparent. Over all, the map has been drawn to a scale of about 1:4,750,000. This is roughly the scale on which Borneo, Sumatra and much of China are drawn. But the scale, it turns out, is not uniform across the map. When you compare the Selden map with the modern conic projection, you can see that some parts are too large and some too small. Measurement bears this out. The Philippines and the northern part of China along the Great Wall have been drawn on a scale that is double the scale of the rest of the map, about 1:2,400,000. This means that these areas are twice the size they should be if they had been properly scaled to Borneo or the rest of China. There are also areas distorted in the other direction. In particular, mainland South-East Asia gets abbreviated. In Yunnan the scale shrinks to 1:6,000,000, and in Vietnam it diminishes to less than 1:7,000,000.

What does this tell us? Does it argue against the idea that our cartographer used a European map as his template to piece together the parts of East Asia? Possibly, but I don't think this is the most interesting conclusion we can derive from the variability of scale. To me it suggests that he was working from another data set. The secret of his data came into view only when the team of conservators – Robert Minte and Marinita Stiglitz of the Bodleian Library and Keisuke Sugiyama of the British Museum – removed the cotton lining to which conservators in an earlier century had glued and varnished it. By 2010 the map was in a sorry state, but the team was able to lift the paper sheet on which the map was drawn without sacrificing any of the original. Once the original had been cleaned and dried, they discovered tell-tale marks on the back: drafts of the scale and rectangle at the top of the map, plus signs that no one can decipher. Far more exciting was the discovery of a series of connected straight-edge lines. When the conservators turned the map back over to compare these with lines on the front, they found a perfect match. The lines on the back exactly replicated the lines that form the main trunk route drawn down the east coast of China on the front. How would this have come about? The obvious explanation is that the cartographer started to draw his map on one side of the paper, then turned the sheet over and redrew it on the other. Maybe he was practising; maybe he realised he had made a mistake. Whatever his reason, he wanted to start again. The important thing here is not that he practised or made a mistake. What the verso lines reveal is the key to his mapmaking method: he drew the sea routes first. Rather than doing as we naturally would, outlining the coasts and then filling in the routes to run between the ports, he drew the route lines first, based on the route data in his rutters, and then filled in the coasts around them. So this map is really not a map at all. It is a chart of sea routes. The landforms are approximate afterthoughts.

The reason for the startling accuracy of the Selden map is now clear. It looks geographically correct not because the mapmaker traced the outline of a European map, although he could well have seen one. It looks as well as it does because it was drawn from the sea. The quality of the result is due to the quality of the data in his rutters. Complete accuracy escaped the Selden cartographer, given his inability to convert time uniformly into distance, but near accuracy was possible. The fact that the Selden and Speed maps resemble each other is hardly surprising, for both patched together this part of the world by working from the water. It could not have been any other way. Both were engaged in roughly the same endeavour: producing charts that would display what Selden called 'the Publique Right of Mutual Commerce' in order to facilitate the movement of cargo vessels through the China seas.

Now for one more startling discovery; at least it startled me. We speculated earlier that the ruler beneath the compass rose may have had something to do with how the map was drawn. Knowing now that the map is scaled at a ratio of 1:4,750,000, can we make the ruler fit with that scale? Let us do the math. If one inch on the Chinese ruler (which as drawn measures 3.75 cm) equals a day's sail at a speed of 6¼ knots, which calculates as 150 nautical miles (240 km), then the distance represented by 1 cm on the ruler is 64 km. This tells us that the ruler was drawn on a ratio of 1:6,400,000. This is too small for the Selden map, although parts of Vietnam were drawn to that scale. But suppose we change the value of an inch on the ruler? One way of getting the scale up to 1:4,750,000 is by slowing down the ships. A slower speed would produce a smaller distance, which would in turn push the scale up to the scale prevailing over most of the map. What speed would do that? Four knots.

This is the startling bit, for 4 knots is the speed that the late Xiang Da came up with when he patiently tried to work out actual distances against reports of time travelled in the rutter. Xiang was the scholar who annotated the Laud rutter for publication in 1959, and so he knew that text better than anyone. If the speed he came up with works out to be the speed that works best the distances on the Selden map, this is hardly coincidental. It can only be because both the Laud rutter and the Selden cartographer were working from roughly the same navigational data.

This was the last discovery I made while writing this book, and it pleased me no end. I have a fondness for Xiang. Xiang was born in 1900, the same year as my mentor in this field, the great Cambridge historian of Chinese science Joseph Needham. In 1935 Xiang travelled from China to Oxford to catalogue the Chinese collection in the Bodleian, just as Michael Shen had done. Recognising the value of the Laud rutter, he copied it out by hand while he was there. He returned to China to serve his country, and ended up being tortured to death at the age of sixty-six during the scandalous first year of the Cultural Revolution for the sin of having travelled abroad and knowing foreigners. I like the fact that Xiang's careful work on the rutter has now come back to prove something about a map he never had the chance to see: that the ruler is indeed the scale by which the cartographer drew the routes on the Selden map. His reputation outlives his tormentors. The vindication is a small one, but among the people I hang out with, this triumph matters. Xiang was one of our finest, and this is what we do.

———————————

We get to the third secret of the map's construction via the one we have just exposed. This secret is that the map has a magnetic signature.

Look carefully at the compass routes. You will recall that most segments of these routes are labelled with their compass bearings, seventy-two of them spaced around the circumference of the compass at an angle of $5°$ from one to the next. The cartographer has done more than mark their bearings. He has gone a step further and drawn the route lines so that they depict the actual angles they are supposed to represent. The reference for these angles is the compass rose. If a line is marked zi ($0°$), then it has been drawn to conform to how zi appears on the compass rose. Now look more closely at the compass, and you will notice that zi, the northern point of the compass, does not in fact point to the top of the map. It is tilted slightly to the left by about 6°. If the rose had been drawn to line up along the map's north-south axis, *zi* should point true north. It doesn't – because it shouldn't.

It is common knowledge that the earth's magnetic field does not align perfectly along its north–south axis, for the magnetic pole is not fixed. It meanders erratically inside the Arctic Circle among the islands of the Canadian north. Actually, as I write, it is just about to leave the sector of the Arctic Ocean under Canadian sovereignty and enter Russia's Arctic sector – claims of jurisdiction, incidentally, that neither country could make were it not for John Selden's argument that it was reasonable to delineate boundaries on the sea. This means that the rate of magnetic declination – the difference between geographical or 'true' north (the top of the map) and magnetic north (the *zi* position on the compass circle) – is always changing. If the Selden compass points 6° to the left, it is because in China at that time the magnetic pole must have been 6° to the left of the North Pole. As it turns out, according to the reconstruction of historic declinations published by the US Geological Survey, 6° to the west is roughly consistent for the eastern edge of Asia in the early decades of the seventeenth century.

The compass rose is not the only thing that tilts. So too do the routes. While she was working on the map, my research assistant, Martha Lee, noticed that the routes did not line up perfectly with the top of the map. In a world without magnetic declination these routes would be drawn in conformity with the Selden map's north–south axis. In fact, they are not. Martha decided to compare the angles at which the lines were drawn with the tilt of the rose. To give the Selden cartographer reasonable room for error, she allowed him a margin of 2½° in either direction, in keeping with the fact that the seventy-two points on the Chinese compass are 5° distant from each other around the circle. The easiest routes to test were those having compass bearings of 180° (*wu*) or 0°/360° (*zi*). Allowing for the margin of error that Martha assigned, they are drawn such that they deviate from the vertical at a rate of roughly 6° to the left. The same pattern is true of the routes on other bearings. The variation from true north is not uniformly 6° to the left in every case, but it is close enough over all to conclude that the route information on the Selden map has been magnetically coded to reflect the position of magnetic north at the time it was drawn. The map, in other words, has a magnetic signature.

Now that we know this secret, there is something else on this map we can decipher, and again I have Martha to thank for the discovery. As she measured the angles of the route lines, she found that the route drawn most accurately in relation to magnetic declination was the coastal freeway – minus the final run to Hirado, which is skewed to the west by a further 5½°. Five other routes were drawn with almost an equal degree of accuracy: the two routes connecting Moon Harbour and Macao to Manila; the route running from Manila down the north-west coast of Borneo, the one that connects the Gulf of Tonkin to Java; and another that runs around Johor to Malacca and up the west side of the Malay Peninsula. The care with which the Selden cartographer drew these routes suggests that he regarded them as the main links that Chinese ships used in the South China Sea network. All the rest drift off their compass bearings. The route from Johor east across the bottom of Borneo to Ternate starts well but soon wanders 10½° off its bearings. The route down the east side of Sumatra to Batavia does the same, although once within range of Batavia it gets back on course. The worst has to be the line running from Moon Harbour out to the Ryukyus and up to Osaka. This route just gets wonkier the further it goes, hitting a degree of error of 16½° off what it should be. But then, all of Japan is wonky: the Selden cartographer had never gone to Japan.

This distortion doesn't mean that the cartographer made a hash of what he was doing. Rather, it signals that, try as he might, he had no hope of ever getting all his routes to line up properly. If we pause to think why this might be, we already know the reason: curvature. Over this great a distance, a course on a straight bearing in reality has to curve; and if you have no way to compensate for the curve – which is what the Mercator projection allows you to do – you end up sailing off at a tangent to the course you intended to travel. Our cartographer didn't know this, nor did he have a projection that could compensate for the three-dimensionality of the globe when he drew his map. He had only one course of action: he had to cheat. To make sure that the most important lines stayed as closely aligned as possible to their true magnetic direction, he had to jimmy the others so that they all linked up plausibly to where they were supposed to. He was probably stumped by his discovery – and he would have had to make it in the course of drawing the map – that uniform accuracy was impossible. The route data were correct, and yet he couldn't combine them visually. Nor did he have a way to account for this difficulty theoretically. The best he could do was control the distortion by cheating here and there on the secondary routes. He did it well, to judge from the end product.

––––––––––

From this revelation flows an even more fascinating discovery leading to the map's fourth secret: there is a pattern to this cheating. This only became apparent when Martha aligned the Selden map to GIS, the Geographic Information System. Using a technique called geo-referencing, she took identifiable points on the Selden map and matched them to their GIS positions. If you think of the Selden map as drawn on rubber, then what she did was stretch it onto a map of the world as we know it today. The process is called 'rubber-sheeting'. A historical map drawn with careful attention to distance and direction will require little stretching when it is rubber-sheeted onto the world. Where the stretch is greatest reveals where the distortion in the original is greatest.

Given the reasonably high accuracy of the map, Martha didn't have to stretch the map much to lay it onto its GIS coordinates. The main thing she ended up doing was to break the original map into three large chunks plus a few smaller pieces (Fig. 26). When she did that, a gap opened in the centre of the map. To put this in terms of what the Selden cartographer did, the only way he could make his map work was to push the land masses around the South China Sea closer to each other than they actually are. The South China Sea was an easy place for him to cheat for there wasn't anything in that sea that anyone wanted to find or get to. Pilots of cargo junks went around it rather than through it. They knew that, otherwise, they would find themselves caught in a dangerous ground of tiny reefs and outcroppings lurking just above or below the water's surface. In the Ming dynasty, and indeed all the way down at least to the nineteenth century, no one knowingly sailed into this zone.

The Selden map does include some islands in the sea, but only those that impinge on the routes running along the coasts. The Pratas Reef off Hong Kong is there, labelled Nan'aoqi (which we might translate as something like 'South-of-Macao Wash'). So too are the Paracels, the Western Shoals, depicted in two sections. The northern Paracels, the Amphitrite Group, have been drawn in the shape of a sail, to which has been added the note 'Sandbars for ten thousand *li* in the shape of a ship's sail'. Just south of this is an island painted red and labelled 'Islet red in colour'. To identify it with any particular island in the Paracels would be to force an interpretation where no further evidence is available. Right below the red island appears a comet's tail of hatchings trailing south. This appears to be the Crescent Group, the southern section of the Paracels. It is labelled 'Reefs for ten thousand *li*': which is to say, a long navigational hazard to be avoided. As for the hundreds of other reefs and atolls over on the east side of the South China Sea, known collectively as the Spratly Islands, they are simply not there. No sea route went through them. Accordingly, it would be tendentious to argue, as some eager nationalists might, that the Selden map proves anyone's claim of sovereignty over any rock in this sea. Declaring sovereignty wasn't what sailors or mapmakers were doing in this part of Asia in the seventeenth century. These were islands nobody wanted.

This realisation leads us back to the purpose guiding the Selden cartographer's hand as he drew the map. This is a commercial navigation chart devoid of imperial designs or claims. Political nations, Ming China included, did not interest our cartographer. But then nor did the sea interest the Ming. The court was not persuaded that it could benefit by allying with commercial interests, as Britain and Holland were doing; the Ming preferred to see itself as presiding over a world order that consisted solely of obedient petitioners to its court. The Cantonese poet Ou Daren expresses this court-centred vision in his poem 'Coming through Plum Pass as the Evening Clears'. Plum Pass is the northern entry-way into Guangdong province through a range of mountains that separates the plains of central China from the south. As Ou clears the pass heading south, he imagines the counterflow of tribute envoys taking the same route north in the direction of the capital:

Sunset lingers on a thousand peaks
Receding into distance as we clear Plum Pass.

The sun settles, the monkeys fall silent,
People head home along the skyline trail.
Going down from the Central Plain, the mountains open a path;
Coming up from the South Sea, the Europeans are monitored.
Here the kings of the world's nations converge;
Here the war horses stand idle in the autumn breeze.

This vision imagines world order as a hierarchy of regulated positions rather than a negotiation of competing claims. It confidently assumes that the authority of the dynasty is sufficient to exert a calming force over troublesome foreigners who, once they have submitted, as a lesser power should, to a greater, approach the throne only to serve. So great are the benefits of their submission that the army can stand down the cavalry. This is the land viewing the water.

The view from the water was quite the opposite. So long as 'the Chineses had refused to trade with the English', as John Saris phrased the situation in his journal, the kings of the world might converge in a complaisant ambassadorial stream just as Ou pictures them doing, but the merchants won't. From where Saris stood on the deck of the *Clove*, war horses were of no account. They couldn't gallop on water, and he was not interested in launching an assault on land. Instead, he would do as he chose with Chinese ships that the indifference of the Ming state left as orphans on the sea. He would board them at will, demand the services of their pilots with impunity and seize their cargo when they resisted. He even petitioned the shogun in Japan for permission to bring captured Chinese cargo ashore to sell; the shogun didn't see this as a good idea. What Saris wanted most was to trade, not to send an embassy, although he would have done so had the opportunity arisen. Prevented from trading, Saris was no different from his Dutch counterparts, who in good Grootian fashion believed that the refusal to trade gave them the right to use force. Selden would have demurred, arguing that the Ming had a right to regulate foreign trade. But Saris was in Asian waters to make money, not to cavil over law.

The effect of Ming policy was a free-for-all on the waters just beyond the reach of naval patrols. The Ming secured its coastal waters in a fashion not unlike the King's Chambers of England, but coastal authority was erratic. When the Wanli emperor signed an edict slashing the funds for 'anti-Japanese defence boats' in 1614, it became even more so. The emperor of 'all within the four seas' was not willing to fund the dominion over the seas adjacent to his realm, in complete contrast to James I – whom Ben Jonson similarly dignified as 'lord of the four seas' – who imagined dominion over his, and could call on John Selden to supply the legal arguments for it.

What the fourth secret, the shrinking of the South China Sea, tells us is that although the Selden map was drawn from the water, it was not drawn to demonstrate any claims of sovereignty over what lay out in the ocean. It was simply a sea chart showing merchants where to go.

The first question anyone would ask about the map is, who drew it? This is a secret we won't be able to unlock. Our cartographer has vanished. The closest we will get is in unravelling two other secrets: where and when he drew it. Let's start with the where. In his will John Selden states unequivocally that his map was 'made there'. The reasonable assumption is that 'there' means China, although how would Selden have known? The language of the labels, as also the use of a prefabricated China map, supposes Chinese authorship, so the cartographer was indeed Chinese, but did he have to be in China to draw it? Is it possible that he was somewhere else?

Consider the places about which he appears to have local knowledge and those of which he doesn't. We can start by ruling out the badly drawn bits, such as Japan, where Nagasaki is Longzishaji – the Langasaque of European lingo. If we move our gaze down to the Philippines, we see that Manila is vividly drawn, plus a whole string of place-names down the west side of Luzon. Luzon is good. However, everything to the south is a complete muddle. These were not routes he knew personally.

To my mind, the part of the map that feels most geographically informed is its southern half. I have already noted that Ming maps abbreviate South-East Asia so radically that it sometimes disappears entirely. The Selden map does South-East Asia like no Chinese map before it and none after it for two centuries. If I had to put the Selden cartographer somewhere, I would put him all the way downstage on Java, either in Bantam or Jakarta. Bantam was the major trading site for Europeans arriving in these waters in the sixteenth century. Chinese knew it as Sunda, the name of the regional sultanate, and this is how it is labelled on the Selden map. Bantam's fortunes shifted in 1609, when a fierce internal struggle drove one leadership faction east, to the nearby town of Jakarta. A decade later the Dutch seized the city and renamed it New Batavia, using the ancient Latin name for the Low Countries. Batavia was used until 1942, when the wartime forces of Japan occupied the city and restored its earlier name of Jakarta. Thomas Hyde has written Nova Batavia beside the place-name label, but the Selden cartographer gives it the term by which Chinese knew it before and after 1619: Jakarta.

Can we do anything with this information? If we turn back to Selden's will, we can extract one more clue about the origin of the map. Selden declares that his map was originally acquired by 'an englishe comander who being pressed exceedingly to restore it at good ransome would not parte with it'. If this story sounds familiar, it is because we have heard it before. At the time I let it pass, but now it is time to return to it. It is the tale Samuel Purchas told of how John Saris acquired the map of China that Purchas obtained and printed in *Purchas his Pilgrimes*. Saris, writes Purchas, confiscated it in Bantam from a Chinese merchant who failed to make good his debts to the East India Company. After Saris returned to England, the Saris map, as I have renamed it, went to the travel writer Richard Hakluyt. After Hakluyt's death in 1616 it passed, along with the rest of his research materials, to Purchas.

Purchas's story about this map is almost identical to Selden's version of how his map was acquired. Are they talking about the same map? That is not possible. The map Purchas reproduced is the standard, squared-off rendering of the country from which the rest of East Asia, except for a prodigious Korea, is absent. The two maps are entirely different. Is it possible that Selden got muddled and attached a story he read in Purchas to his own map? There is no reason to think so. If the stories echo one another, it may be because they are two parts of the same story.

The will provides yet another clue. It says that the map came from 'an englishe comander'. The term is a technical one. Each expedition the East India Company sent to Asia was called a 'voyage', later a 'joint stock voyage', and each was led by a 'commander'. The question to which the will gives rise is, which of them got back to London and passed this map to Hakluyt or Purchas? The candidates are few enough. Anthony Hippon, commander of the seventh voyage, died in Pattani in 1612. The famous commander of the sixth, Henry Middleton, died the following year in Bantam, as may have the commander of the fourth, Alexander Sharpleigh. Henry Middleton's brother David, who commanded the fifth voyage and the third joint stock voyage, drowned when his ship went down in a storm off the coast of Madagascar in 1615. Nicholas Downton died in Bantam in 1615. Of the commanders who managed to return to England, James Lancaster died in London in 1618, William Keeling on the Isle of Wight in 1620, and Martin Pring in Bristol in 1626. Thomas Best is the most elusive of this crowd. Famed for defeating a Dutch fleet off the coast of India during the tenth joint stock voyage, he resigned from the EIC in 1617 and died in Stepney in 1639.

None of these men has left any evidence of having acquired carto-graphic documents in Asia. The process of elimination leaves the 'coman-der' of the eighth voyage – spelled in Company documents exactly as it appears in Selden's will: John Saris. Spelling doesn't tell us anything, but it did catch my eye while I was leafing through Company minutes. Once I noticed the coincidence, the little scraps of evidence I had been collect-ing suddenly pointed in Saris's direction. It is not necessary that Selden and Saris met. Selden probably got the map from Samuel Purchas via Richard Hakluyt. This was a route by which other items in his library came to him, most famously the illustrated album of the life and history of the Aztec people, known as the Codex Mendoza. Commissioned in the 1540s for the king of Spain, the codex was seized by French priva-teers, obtained by the king's cosmographer André Thevet, then sold to Richard Hakluyt. After his death in 1616 it went to Purchas. Ten years later it passed from Purchas's estate to Selden, who duly inscribed it with his motto, *peri pantos ten eleutherian* ('Above All, Liberty'). It too resides in the Bodleian Library. The same chain of collectors, I now believe, conveyed the map from Saris to Selden.

If I am right, then the odds are good that Saris acquired it from a Chinese merchant in Bantam. There the trail goes cold. Whether it was originally drawn in Bantam is anyone's guess, although I am tempted to think it was, commissioned at no little expense by a Chinese merchant who ran his trading operations out of this port and desired nothing more than to see his commercial empire displayed on his wall. Our hypothetical owner is as far as we can go. Unless another map in the Selden cartographer's distinctive hand and style surfaces, its maker must remain anonymous.

This does not mean that we can't know a great deal about him. He was Chinese. He had access to sources about Chinese sea routes, including a rutter dating back to the fifteenth-century voyages of Zheng He. He was involved more with the south than the north, mainly the Western Sea route, but he had knowledge of the Northern and Eastern Sea routes as well. His grasp of the spatial relationships among the places he maps is so good that it is hard to believe that he himself did not sail the South China Sea. How else could he have had the confidence to produce such a perfect mindscape of the Chinese trading world into which the Europeans had come?

So we can say that we do know him after all, even if we can never put a name to him. That must remain a piece of the puzzle we will never find. And even if we can't know his name and can't quite determine where he drew the map, we can figure out when. We have already come close.

There were three times when John Saris could have acquired it. The third and last time was his final call into Bantam, in early 1614, on his way home from Japan. This doesn't seem the likely occasion. He was there for a little over five weeks, and while he did have dealings with Chinese merchants during his stay, he mentions nothing in his journal about dunning them for unpaid debts. Besides, he was a commander, not a bailiff. Collecting debts would have been beneath the dignity of his position. The second time was when he arrived on his way to the Spice Islands before heading to Japan. He was in port for two and a half months in the winter of 1612–13, but during that time was much occupied with loading and despatching two of his ships back to London. It is equally hard to imagine him out debt-collecting.

That leaves his first stint in Bantam, a five-year posting that started in 1604, culminated with his promotion to the post of Chief Factor in 1608 and ended with his departure home on 4 October 1609 to lobby for promotion to commander. This strikes me as the likeliest time for acquiring the map. The job of Chief Factor would certainly have included collecting bad debts from Chinese merchants with whom the Company had done business. If Selden got the story right, the merchant from whom Saris took the map 'pressed' him 'exceedingly to restore it at good ransome'. One can imagine all the reasons, from the cost that must have gone into producing the map to the value of the trading knowledge it recorded, especially for foreign newcomers unfamiliar with the trade routes of the region, to the political liability of having let strategic information about China fall into the hands of a foreigner. Foreigners were not allowed to have maps of China, as I learned centuries later.

If 1609 was the date by which Saris confiscated the map, we can also determine the date before which it could not have been made. The map in fact has a very precise time signature, buried in the label beside Wanlaogao that so intrigued Thomas Hyde: 'Where Red Hairs live'. The label, as we know, refers to the founding of the first Dutch fort on Ternate in May 1607, which resulted in the famous division of the island between them and the shapeshifting Spanish that caught the attention of our armchair navigator, Zhang Xie. The earliest possible date for the label is thus 1607. So now we can match the *terminus ante quem* of 1609 with a *terminus post quem* of 1607. I suggest we split the difference. Circa 1608, then. We have a date.

The search for the Selden map has been more convoluted and complicated than I expected it to be when I started out, a circling maze rather than a straight path: from John Selden forward to Thomas Hyde and Michael Shen; from John Selden back through Samuel Purchas and Richard Hakluyt to John Saris; from the map itself forward to Zhang Xie and his study of routes, to William Laud and the rutter he could not decipher; from the map back to Zhang Huang and his encyclopaedia, and even back beyond him to Luo Hongxian and his national atlas; with a sideways look from Selden to James I and the divine right of kings to connect him to the law of the sea in the first place; plus John Speed and his world atlas thrown in to give us a point of comparison.

But we have got there, to the origin of the map – well, more or less. With an unsigned and undated document such as ours, this is not a bad showing. Many a historical expedition gets lost at sea before reaching its destination. Perhaps ours hasn't quite made it to home port, but at the very least we have achieved something: we have given back to the Selden map some of the history it lost centuries ago. And not just that; for in giving the map its history, we have written ourselves into the story.

Epilogue:
Resting Places

During the first half of the seventeenth century the Selden map was the most accurate chart of the South China Sea in existence. There had never been a better map, and for another four decades there wouldn't be another. Yet from the larger perspective of the history of cartography, the Selden map never got the chance to make a difference. The man who designed the map came up with an ingenious method for picturing the world by working from the sea and not from the land. It was a smart hunch that got him half-way past the problem of curvature and resulted in a stunning map. But unless or until we discover other maps like it, we have nothing else to go on. Could this be a one-off, a brilliant exercise that produced one map and no other? Hard to imagine, and yet no other exists. The cartographer left no notes explaining how he drew it, and he taught no students who could go on to refine and generalise his method into a set of principles for drawing maps on this scale. Nothing was learned or passed on, so far as the surviving evidence shows. It was a dead end.

The map's transfer to Europe could have altered this story. Shown to the right people, the map might have had an impact on European cartographers. But it didn't. Richard Hakluyt and Samuel Purchas saw it, yet there is no evidence that they did anything with it. By the time the map was on display in Oxford, it was too late to make a difference. Other developments intervened. We can date the leaching of the map's value quite precisely to the year 1640, by which date the great Amsterdam cartographer Joan Blaeu – who made those globes James II pointed to in the Bodleian – produced a portolan chart of the China seas for the Dutch East India Company of unsurpassed accuracy. This chart marks the moment after which Europeans could count on their own version of the China seas. There was no longer a need for anyone to go back to the Selden map as a cartographic source. As of 1640, the Selden map had to drop back from the front line of mapmaking. When the eminent scientist Edmond Halley saw it in 1705, he dismissed it as inaccurate.

This realisation leaves us with a split verdict on John Selden's role. On the one hand, we have him to thank for preserving the map. Without his passion for collecting Asian manuscripts, we would not have known that such a map could even exist. By sequestering it at home, however, Selden effectively withdrew it from circulation. He showed it to friends during the quarter-century when it was the finest map of Asia. But he didn't put it in the hands of the geographers who might have learned something from it. This map could have directed Europeans in their efforts to improve their image of the world, but it didn't. By the time it reached the Bodleian Library, it had nothing left to tell cartographers that they didn't already know, and know in what they regarded as a technically superior way. The map's intuitive accuracy was now purely of historical interest. Better methods were available for mapping Asia. For the students of Oxford it became little more than a foreign curiosity that no longer mattered. Its fate was to be forgotten. Eventually it was taken down, rolled up and laid to rest in the bowels of the Bodleian, to hibernate there until such time as we were prepared to recognise its value at the far end of the corridor that runs from the heady age when it was made, four centuries ago, to our own madly globalising present. It had to wait for us to catch on, and catch up.

The map did enjoy a curious afterlife, however, by virtue of the company it was forced to keep.

Thomas Bodley's new library in Oxford occupied the upper floor of the Divinity School. A decade after it opened, that building was extended, and a dozen schools (which in universities today would be called faculties) were set up in suites of rooms around what came to be known as the Schools Quadrangle. The Anatomy School occupied the south-west corner. This position meant that it and the Bodleian shared the same entrance: books up and bodies down.

Over time the Bodleian Library found itself the unsuspecting recipient of curiosities that English travellers picked up abroad and thought deserved preservation in a public institution. Thus it was that the Bodleian came to own a crocodile that Oliver Cromwell's brother-in-law bagged in Jamaica in 1658, a 'sea elephant' purchased in 1679, a mummy acquired by a merchant in Turkey in 1681 and the desiccated body of an African boy that entered the collection in 1684, among other oddities. Collecting natural curiosities may not have been quite what Bodley had in mind when he revived the university library, but this became the passion of the century. In they came, and Hyde as Keeper had to put them somewhere. After 1683 he tended to send natural specimens over to the new museum that Elias Ashmole opened that year, known as the Ashmolean. Human specimens he sent downstairs to the Anatomy School. Apparently for want of a better location for its display, this is where the Selden map found its eighteenth-century resting place: on a wall by a staircase in the Anatomy School.

The map did not hang alone, for which we may have Thomas Hyde to thank or blame, depending on how you look at it. This story takes a bit of telling.

Five years after Michael Shen left Oxford, another Asian arrived. He was a Pacific Islander by the name of Giolo. 'This famous painted Prince is the just wonder of the Age', proclaims a London handbill inviting the public to view this human curiosity. 'His whole Body (except Face, Hands and Feet) is curiously and most exquisitely painted or stained full of Variety of Invention, with prodigious Art and Skill perform'd.' The 'noble Mystery' of tattooing has never offered a display as fine as what you can

see in 'this one stately Piece' (Fig. 27). The handbill insists that tattoos
are not mere decoration but a badge of royalty and a prophylactic against
venom. When a prince has been newly tattooed, he is carried naked into
a room with poisonous snakes and insects while the king and his entire
court gather to watch. If the tattoos have been properly applied, the creatures adore the prince and do him no harm. Even more remarkable for
what concerns us is that his 'more admirable Back-parts' are tattooed
with 'a lively Representation of one quarter part of the World upon and
betwixt his shoulders', from 'the Arctick and Tropick Circles' between
his shoulder blades to 'the North Pole on his Neck'. So Prince Giolo was
a walking north-conical Mercator projection, or that's what his promoters seemed to think he was. Those desiring to discover 'what Wisdom
and ancient Learning may lie veiled under those other curious Figures
and mysterious Characters scattered up and down his Body' were invited
to view the prince at the Blue Boar Inn on Fleet Street. There was also a
business angle, for the handbill declares that Giolo comes from 'a fruitful
Island abounding with rich Spices and other valuable Commodities'. He
is 'neat and cleanly', but no one understands his language and he doesn't
speak a word of English.

Giolo's instant celebrity was enough to gain him an audience with
the king. It also inspired an instant book. The title-page of *An Account
of the Famous Prince Giolo* promises to reveal everything concerning 'his
Life, Parentage, and his Strange and Wonderful Adventures; the Man-
ner of his being brought for England: with a Description of the Island
of Gilolo, and the Adjacent Isle of Celebes: their Religion and Manners'. It is a short book. The first eight pages offer a scholarly account
of the indigenous religion of the Celebes (now known as Sulawesi). The
narrative then switches to a rousing tale of Giolo's exploits to rescue
his beloved Princess Terhenahete from her cruel captors. Among the
perfidious characters attempting to foil the hero is the pilot of the ship
on which the lovers escape with her father. Pilots make good suspects,
for no one likes placing complete trust in a total stranger. Giolo's tragic
mistake is to spend the night with Terhenahete below decks 'in all the
Joyes of Love' when he should have been above decks scanning the seas
and directing their course. The pilot betrays them by sailing the ship
back to where they started. As our heroes are about to be captured, the

princess's father stabs the pilot for his treachery and throws his corpse overboard.

It is good entertainment, and absolutely none of it is true. The man who later contradicted the hype was none other than the English sailor who brought Giolo to England, the privateer William Dampier (Fig. 28). In his popular travel memoir Dampier calls him Jeoly. He was a slave Dampier acquired while he was on the southern Philippine island of Mindanao as part of a property settlement. Giolo had indeed been captured and enslaved, but by Moro traffickers operating out of Mindanao, not by a rival royal lineage. So, no royal pedigree, no Princess Terhenahete, no slain pilot and no home in the Spice Islands. The man was from Miangis, a tiny Pacific island well east of Mindanao.*

When they got to England, Dampier says he lost control of his share in human property to the ship's mate, who was also part-owner of Giolo. 'I was no sooner arrived in the Thames, but he was sent ashore to be seen by some eminent persons', Dampier writes. 'I being in want of Money, was prevailed to sell first, part of my share in him, and by degrees all of it. After this I heard that he was carried about to be shown as a Sight.' As for the story of the snake pit, Dampier scoffed, 'I have seen him as much afraid of Snakes, Scorpions, or Centapees, as my self'. He says nothing about a map tattoo.

An Account of the Famous Prince Giolo has been attributed to none other than Thomas Hyde. The evidence is the pseudo-anthropological account of the religion of the Celebes, which few besides Hyde had the knowledge to manufacture; also that Giolo went next to Oxford after his London showings. Also it would make sense that Hyde was interested in Giolo, for he had been looking for a native speaker of Malay. Giolo did not

---

* Miangis got into the news in the 1920s, when the United States and the Netherlands went to the Permanent Court of Arbitration in The Hague to argue over who owned the island. The Americans had dominion over the Philippines and judged that Miangis, 60 kilometres from Mindanao, belonged to them. The Dutch argued that it was part of the Dutch East Indies. The Swiss arbitrator found in favour of the Dutch, on the grounds that they had occupied the islands of this region since the seventeenth century and had exercised 'peaceful and continuous display of State authority over the island', and that the island was in any case, as he put it, 'inhabited only by natives'. Indonesia inherited the island after the Second World War and retains control today, a legacy of Dutch imperialism.

speak Malay, as it turned out, but Hyde would not have learned that until he had met him. Still, would England's leading authority on ancient Asian religion really have been willing to get involved in this sort of sensation? Perhaps it was the prospect of making easy money. Some doubt the attribution. But with someone as evasive as Hyde, alas, you never know.

There is no record of Giolo's sojourn in Oxford except that he died there of smallpox. The university ordered him to be buried in the churchyard at St Ebbe's, on the west side of the city. At that point the historical record runs cold. He is not listed in the parish register, and when I went hunting for evidence at St Ebbe's for a gravestone, I discovered not only that the original church had been demolished and rebuilt in 1813, but that most of the burial ground has since disappeared beneath the Westgate Shopping Centre. Some ten tottering headstones decorate the garden now, and none is Giolo's. But like as not he wasn't given one. Churchyards were for Christians.

Giolo did not entirely disappear, however. On his death, it was decided that his tattoo was too good to lose. In the scientific climate of the time, Giolo's skin was considered useful for advancing knowledge. In order to preserve this asset, the university appointed Theophilus Poynter to flay the corpse. Poynter taught in the Anatomy School and was Oxford's most successful surgeon. Anthony Wood went to him, and Hyde probably did so as well. He taught downstairs, and his private surgery was in Cat (now Catte) Street, just down the lane from the Bodleian. According to a catalogue of objects in the Anatomy School's collection compiled in 1709, the year Poynter died, he had skinned others. On the list is a skeleton Poynter had arranged 'according to the natural motion' along with 'the skin taken from it, whereon is the hair and nails'.* These were the remains of a convict executed in London and shipped to Oxford for the still sensitive practice of medical dissection. As the anatomists after 1683 did their work next door in the basement of the Ashmolean Museum, now the Museum of the History of Science, this must have been where Poynter flayed Giolo.

---

* These were not the only human remains in the collection. Among other items the catalogue lists the skeleton of a woman who was hanged for killing four of her eighteen husbands. A German visitor the following year reports that the School had her 'stuffed Skin' as well.

We do not know who decided, or why, that Giolo's skin should hang on the wall next to the map of China. But there they were for some years, Giolo's tattoo and Selden's map, strange wallfellows paired as Asian curiosities for the edification of scientific visitors. If there ever was a map of the Pacific Ocean on Giolo's back, we have no way of checking. The Selden map survived, but Giolo's skin has long since vanished.

———————

John Selden never went to sea. Famed in his own lifetime as 'the Glory of our Nation for Orientall learning', he never travelled further than Oxford. He died in 1654 at the home Elizabeth Talbot left him in Whitefriars, and his body was taken two blocks west to be buried in the Temple, the thirteenth-century church of the Templars to which his law society, the Inner Temple, was attached. It was the suitable place for his tomb, for the Inns of Court was where he had come of intellectual age and built a reputation that even he could not have predicted as the greatest constitutional lawyer of the age. His body was laid beneath the floor, a stone set over him, and a black marble tablet placed on the wall to commemorate his life.

Not a few have wondered whether his executors did as well by the great man as he deserved. When Samuel Pepys visited the Temple Church on 22 November 1667, which had been spared in the Great Fire of the previous year, he was dismayed by 'the plainness of Selden's tomb, and how much better one of his executors hath, who is buried by him'. The tomb that offended Pepys was that of Rowland Jewkes, installed two years before Pepys's visit. Selden and Jewkes met in 1621, when the younger man entered the Inner Temple, and the two became fast friends. Pepys's mildly malicious comment implies that the executors should have built Selden a better memorial than the one Jewkes arranged for himself. Having gained financially from the will, Jewkes might not have enjoyed such a fine monument had he not been a beneficiary of Selden's estate.

In its section on the Temple, a London guidebook of 1708 indirectly confirms Pepys's judgement. Jewkes's tomb, listed as no. 32 in the roster of monuments to see in the Temple, is described as 'a handsome white marble Mon[ument] adorned with Col[oured] Entablature, &c. of the

Ionic Order, enr'ched with Cherubims, Festoons, Urn, &c'. The Latin inscription identifies Jewkes as 'one of four executors of the will of the great Selden', as though that were enough to ensure immortal fame. To its left is no. 31, described merely as 'a Monument in memory of John Selden, dated 1654'. No cherubims and festoons for Selden. The guide-book writer was underwhelmed.

In the race for eternity, though, Jewkes lost. His monument is no lon-ger there. If it survived until 1940, the bombing of London during the Blitz pulverised it. The one memorial that remained unscathed is the stone laid in the floor over the spot where Selden is buried. It is still there, in fact, in the south aisle. Centuries of renovations have meant that the floor is now a foot higher than where it was when Selden was laid to rest. His gravestone today lurks inside a murky cavity covered by a scuffed sheet of Perspex.

If popular history were permitted to judge whose graves should be preserved and whose allowed to disappear, Selden's would prob-ably have gone the way of the others lost in the Blitz. A fellow historian declared to him in a letter that 'it were an Ignorance beyond Barbarism not to know you', but most of us have never heard of him. Until David Helliwell showed me the map, I was numbered among the Barbarians, as were you too, I suspect. There is much for which he deserves to be remembered. Some people make a difference to the era in which they find themselves, some to the ages that come after; John Selden, to both. Perhaps his map of China counts as the least of his claims on us, yet it may be the break he needed to keep his memory from sinking even deeper into the foundations of the Temple Church.

A light installed beneath the floor illuminates the sunken stone. If the custodian has neglected to turn it on, the switch is on the pillar behind you.

Appendix I
Boxing the Chinese Compass

degrees	Chinese name		meaning	direction
0°	子	zi	Earthly Branch 1	north
5°	癸子	guizi	zi by gui	
10°	子癸	zigui	gui by zi	
15°	癸	gui	Heavenly Stem 10	
20°	丑癸	chougui	gui by chou	
25°	癸丑	guichou	chou by gui	
30°	丑	chou	Earthly Branch 2	
35°	艮丑	genchou	chou by gen	
40°	丑艮	chougen	gen by chou	
45°	艮	gen	Trigram 7: Mountain	north-east
50°	寅艮	yingen	gen by yin	
55°	艮寅	genyin	yin by gen	
60°	寅	yin	Earthly Branch 3	
65°	甲寅	jiayin	yin by jia	
70°	寅甲	yinjia	jia by yin	
75°	甲	jia	Heavenly Stem 1	
80°	卯甲	maojia	jia by mao	
85°	甲卯	jiamao	mao by jia	
90°	卯	mao	Earthly Branch 4	east
95°	乙卯	yimao	mao by yi	
100°	卯乙	maoyi	yi by mao	
105°	乙	yi	Heavenly Stem 2	
110°	辰乙	chenyi	yi by chen	

degrees	Chinese name		meaning	direction
115⁰	乙辰	*yichen*	*chen* by *yi*	
120⁰	辰	*chen*	Earthly Branch 5	
125⁰	巽辰	*xunchen*	*chen* by *xun*	
130⁰	辰巽	*chenxun*	*xun* by *chen*	
135⁰	巽	*xun*	Trigram 5: Wind	south-east
140⁰	巳巽	*sixun*	*xun* by *si*	
145⁰	巽巳	*xunsi*	*si* by *xun*	
150⁰	巳	*si*	Earthly Branch 6	
155⁰	丙巳	*bingsi*	*si* by *bing*	
160⁰	巳丙	*sibing*	*bing* by *si*	
165⁰	丙	*bing*	Heavenly Stem 3	
170⁰	午丙	*wubing*	*bing* by *wu*	
175⁰	丙午	*bingwu*	*wu* by *bing*	
180⁰	午	*wu*	Earthly Branch 7	south
185⁰	丁午	*dingwu*	*wu* by *ding*	
190⁰	午丁	*wuding*	*ding* by *wu*	
195⁰	丁	*ding*	Heavenly Stem 4	
200⁰	未丁	*weiding*	*ding* by *wei*	
205⁰	丁未	*dingwei*	*wei* by *ding*	
210⁰	未	*wei*	Earthly Branch 8	
215⁰	坤未	*kunwei*	*wei* by *kun*	
220⁰	未坤	*weikun*	*kun* by *wei*	
225⁰	坤	*kun*	Trigram 2: Earth	south-west
230⁰	申坤	*shenkun*	*kun* by *shen*	
235⁰	坤申	*kunshen*	*shen* by *kun*	
240⁰	申	*shen*	Earthly Branch 9	
245⁰	庚申	*gengshen*	*shen* by *geng*	
250⁰	申庚	*shengeng*	*geng* by *shen*	
255⁰	庚	*geng*	Heavenly Stem 7	
260⁰	酉庚	*yougeng*	*geng* by *you*	
265⁰	庚酉	*gengyou*	*you* by *geng*	
270⁰	酉	*you*	Earthly Branch 10	west
275⁰	辛酉	*xinyou*	*you* by *xin*	
280⁰	酉辛	*youxin*	*xin* by *you*	
285⁰	辛	*xin*	Heavenly Stem 8	

degrees	Chinese name		meaning	direction
290°	戌辛	xuxin	xin by xu	
295°	辛戌	xinxu	xu by xin	
300°	戌	xu	Earthly Branch 11	
305°	乾戌	qianxu	xu by qian	
310°	戌乾	xuqian	qian by xu	
315°	乾	qian	Trigram 1: Heaven	north-west
320°	亥乾	haiqian	qian by hai	
325°	乾亥	qianhai	hai by qian	
330°	亥	hai	Earthly Branch 12	
335°	壬亥	renhai	hai by ren	
340°	亥壬	hairen	ren by hai	
345°	壬	ren	Heavenly Stem 9	
350°	子壬	ziren	ren by zi	
355°	壬子	renzi	zi by ren	
360°	子	zi	Earthly Branch 1	north

Appendix II
Coast Comparison

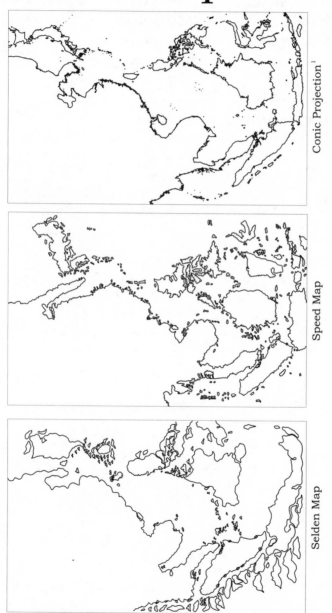

Conic Projection[1]

Speed Map

Selden Map

Acknowledgements
and Sources

Of many who made possible the writing of this book, I must begin with my old friend David Helliwell, who first alerted me to the map's existence and assisted me unstintingly at every stage of research. Next must come my newer friend Will Poole, who was endlessly generous in sharing his vast, intimate knowledge of seventeenth-century Oxford: I could never have made this book as much about England as about China without his guidance. I wish to acknowledge as well Robert Batchelor for discovering the map and sharing his insights on it; also Robert Minte and Marinita Stiglitz, who were most gracious in patiently explaining to me what they learned during the process of restoring the map. I am also grateful to Martha Lee for geo-referencing the Selden map and showing me how it 'worked'.

I have not met or corresponded with Gerald Toomer, and he may not endorse my version of John Selden, but I would like to acknowledge the impact on me of his work on the history of English Orientalism and of his enthralling intellectual biography of Selden. Without his research I could never have wandered so far from my familiar zone of knowledge.

I owe much to those who helped to bring the book into print: my indefatigable agent, Beverley Slopen; my three publishers, Andrew Franklin at Profile, Peter Ginna at Bloomsbury and Sarah MacLachlan at Anansi; and my two editors, Penny Daniel and Janie Yoon. All six know how much they did to shape the book. My only regret is that Peter Carson, my original editor, who was anxious to acquire the book for Profile, died before he could see the final version. His claim that he liked the first draft did more than he could know to buoy me through the months of rewriting. I hope that in the process of revising I didn't misplace the book he wanted.

Jim Wilkerson, Brantly Womack, Richard Unger, Paul Eprile and

Aaron Rynd read part or all of the manuscript and saved me from numerous factual errors and stylistic flaws. As for what I know about sailing, I have Keith Benson to thank. My final thanks, as always, are for Fay Sims, who asked the right questions and helped me to find my way to the end of this book.

The writing of this book was supported in part by a research grant from the Social Sciences and Humanities Research Council of Canada.

Preface
The Waldseemüller map is presented in John W. Hessler and Chet Van Duzer, *Seeing the World Anew* (Washington, DC: Library of Congress, 2012). For an account of the map's provenance and significance, see Jerry Brotton, *A History of the World in Twelve Maps* (London: Allen Lane, 2012), pp. 146–85.

The technical details of the Selden map have been taken from Robert Minte and Marinita Stiglitz, 'Conservation of the Selden Map of China', unpublished paper delivered at the Selden Map of China Colloquium, Bodleian Library, 15 September 2011.

1. What's Wrong with this Map?
The 2001 air collision off Hainan Island is described in Bill Turnbull, 'Looking at a Miracle', *Naval Aviation News* (September–October 2003), pp. 20–23. On the law of the sea I have relied principally on Donald Rothwell and Tim Stephens, *The International Law of the Sea* (Oxford: Hart Publishing, 2010); for the overflight of military aircraft, see pp. 282–4. I am grateful to my neighbour Michael Byers for his help on issues of the law of the sea.

'Mindscape': Cordell Yee, 'Chinese Cartography among the Arts: Objectivity, Subjectivity, Representation', in *The History of Cartography*, ed. J. B. Harley and David Woodward (Chicago: University of Chicago Press, 1994), vol. II, bk 2, pp. 128–69.

On the islands in the South China Sea, see David Hancox and Victor Prescott, 'A Geographical Description of the Spratly Islands and an Account of the Hydrographic Survey amongst Those Islands', *Maritime Briefing* 1:6 (1995); Jeannette Greenfield, 'China and the Law of the Sea', in *The Law of the Sea in the Asian Pacific Region: Development and*

Prospects, ed. James Crawford and Donald Rothwell (Dordrecht: Martinus Nijhoff, 1995), pp. 21–40; Brantly Womack, 'The Spratlys: From Dangerous Ground to Apple of Discord', *Contemporary South East Asia* 33:3 (2011), pp. 370–87; Clive Schofield et al., 'From Disputed Waters to Seas of Opportunity: Overcoming Barriers to Maritime Cooperation in East and Southeast Asia', National Bureau of Asian Research Special Report 30 (2011). I am grateful to Brantly Womack for providing me with his insights and some of these materials.

Ten thousand European trading vessels: Basil Guy, *The French Image of China before and after Voltaire* (Geneva: Institut et Musée Voltaire, 1963), p. 31.

Chinese mariners sailing 'as securely as the Portuguese': Louis Lecomte, *Memoirs and Observations Topographical, Physical, Mathematical, Mechanical, Natural, Civil and Ecclesiastical, Made in a Late Journey through the Empire of China* (London, 1696), p. 230, quoted in Joseph Needham, *Science and Civilisation in China*, vol. IV, pt 3 (Cambridge: Cambridge University Press, 1971), p. 379.

2. Closing the Sea

On Selden I have relied on the two-volume biography by G. R. Toomer, *John Selden: A Life in Scholarship* (Oxford: Oxford University Press, 2009). His earlier book, *Eastern Wisedome and Learning: The Study of Arabic in Seventeenth-Century England* (Oxford: Clarendon Press, 1996), provides the context of Oriental studies within which Selden worked. Other works on Selden that I have consulted include: John Barbour, *John Selden: Measure of the Holy Commonwealth in Seventeenth-Century England* (Toronto: University of Toronto Press, 2003); Paul Christianson, *Discourse on History, Law, and Government in the Public Career of John Selden, 1610–1635* (Toronto: University of Toronto Press, 1996); and Jason Rosenblatt, *Renaissance England's Chief Rabbi: John Selden* (Oxford: Oxford University Press, 2008). John Selden's will is preserved in the Public Record Office at Kew.

For Anthony Wood's experience at the Bodleian, see *The Life and Times of Anthony Wood, Antiquary, of Oxford, 1632–1695*, ed. Andrew Clark (Oxford: Clarendon Press, 1891–1900); the story of the spectacles appears in vol. I, p. 282. The disposition of Selden's library is discussed in Macray, *Annals of the Bodleian Library*, pp. 77–86; the costs are noted on p. 86. See also D. M. Barrett, 'The Library of John Selden and its Later History', *Bodleian Library Record* 3:31 (March 1951), pp. 128–42. The original list of Selden's manuscripts drawn up after his death, which appears as Appendix C of this article, includes six copies of the Magna Carta, three bound in leather (ibid., pp. 257, 264–5).

'None should be admitted of this House': Edward Hutton, *A New View of London: or, An Ample Account of that City, in Eight Sections* (London: John Nicholson and Robert Knaplock, 1708), II, p. 693. I am grateful to Paul Cleaver for sharing this absorbing book with me when I visited him in Puerto Escondido.

On Ben Jonson, I benefited from reading David Riggs, *Ben Jonson: A Life* (Cambridge, MA: Harvard University Press, 1989), and Robert Evans, *Ben Jonson and the Poetics of Patronage* (Lewisburg, PA: Bucknell University Press, 1989). The quotes from the masques are taken from his *Masques and Entertainments*, ed. Henry Morley (London: George Routledge, 1890), pp. 220, 259, 261–2, 290, 428. His poems are quoted from *The Poetical Works of Ben Jonson*, ed. Robert Bell (London: John Parker, 1856), p. 217. The comment by William Drummond appears in *Notes of Ben Jonson's Conversations with William Drummond*, ed. David Laing (London: The Shakespeare Society, 1842), p. 40. 'Of all actions of a man's life': *The Table Talk of John Selden*, ed. Richard Millwood (Chiswick: Whittingham, 1818), p. 90; 'to please himself', p. 118. Ian Donaldson, *Ben Jonson: A Life* (Oxford: Oxford University, 2011), p. 363, doubts that Jonson accompanied Selden to his first meeting with the king.

On de Groot in the context of Dutch interests in Asia, see Martina Julia van Ittersum, *Profit and Principle: Hugo Grotius, Natural Rights Theories and the Rise of Dutch Power in the East Indies (1595–1615)* (Leiden: Brill, 2006), especially pp. 1–59; and Peter Borschberg, *Hugo Grotius, the Portuguese and Free Trade in the East Indies* (Singapore: National University of Singapore Press, 2011), pp. 41–94. Quotations from de Groot are taken from Hugo Grotius, *The Freedom of the Seas*, ed. Ralph Magoffin (New York: Oxford University Press, 1916) pp. 20, 49.

For Selden's *Mare Clausum* I have used the English translation, *Of the Dominion, or, Ownership of the Sea*, trans. Marchamont Nedham

(London, 1652). His praise of de Groot as 'a man of great learning' appears in bk 1, ch. 26. On Selden's role in shaping the law of the sea, see Thomas Fulton, *The Sovereignty of the Sea* (Edinburgh: William Blackwood, 1911), pp. 338–71; for Charles's call for 'some public writing', see p. 364.

Letter from a colleague (Nicolas-Claude Fabri de Peiresc) in Paris: 'Selden Correspondence', Bodleian Library, Selden supra 108, p. 50; I am indebted to Will Poole for making his transcription of the letters available to me. 'In a troubled state': *The Table Talk of John Selden*, p. 125; 'is dangerous because we know not where it will stay', p. 149; 'every law is a contract', p. 74.

The description of Bartholomew Fair is taken from an anonymous pamphlet, *Bartholomew Faire, or Variety of Fancies* (London, 1641; reprinted London: John Tuckett, 1868), p. 1. Selden's letter to Jonson on the subject of cross-dressing is translated in Jason Rosenblatt, *Renaissance England's Chief Rabbi: John Selden*, pp. 279–90.

'Quod Seldenus nescit, nemo scit': letter from James Howell to John Selden, Selden Correspondence, Bodleian Library, Selden supra 108, p. 218.

3. Reading Chinese in Oxford

James II's visit to the Bodleian Library is recorded in *The Life and Times of Anthony Wood*, vol. III, p. 235. This incident has been retold several times, most recently in Nicholas Dew, *Orientalism in Louis XIV's France* (Oxford: Oxford University Press, 2009), pp. 205–6. On Charles II touching for the King's Evil, see Jenny Uglow, *A Gambling Man: Charles II's Restoration Game* (New York: Farrar, Straus and Giroux, 2009), pp. 54–5; for an instance in Oxford in 1663, see *The Life and Times of Anthony Wood*, vol. I, p. 497.

The two globes in the Bodleian were manufactured by the Blaeu family in Amsterdam (first edition published in 1616) and presented in 1657; I am grateful to Will Poole for the identification.

On Michael Shen, see Theodore Foss, 'The European Sojourn of Philippe Couplet and Michael Shen Fuzong, 1683–1692', in *Philippe Couplet, S.J. (1623–1694): The Man Who Brought China to Europe*, ed. Jerome Heyndrickx (Nettetal: Steyler Verlag, 1990), pp. 121–42. On his training in European languages, see Pierre Rainssant's letter to Pierre

Bayle, 19 March 1686, on-line @ http://bayle-correspondance.univ-st-etienne.fr/?Lettre-540-Pierre-Rainssant-a.

Couplet is the subject of Heyndrickx, *Philippe Couplet*. His approach to Joan Blaeu and his possession of the Luo Hongxian atlas are mentioned in C. Koeman, *Joan Blaeu and his Grand Atlas* (Amsterdam: Theatrum Orbis Terrarum, 1970), pp. 85, 88.

'The Chinese, from the beginning of their origin': *The Morals of Confucius* (London: Randal Taylor, 1691), pp. xvii, xix. This little book was a popular version of *Confucius*, first published in French in 1687 and in English four years later. The text is by Couplet.

The Bodleian's acquisition of *Confucius Sinarum Philosophus* is noted in Macray, *Annals of the Bodleian*, p. 76; for James I's dismissal of the notion that Mary was immaculately conceived, see p. 26. The request from the astronomy professor (Edward Bernard) for a copy of the book from his friend Thomas Smith is mentioned in Dew, *Orientalism in Louis XIV's France*, p. 206, n. 3. On the Jesuit apprehension of Confucius, see Lionel Jensen, *Manufacturing Confucianism* (Durham: Duke University Press, 1997), 118–25.

The denunciation of Brian Walton is mentioned in Toomer, *Eastern Wisedome and Learning*, p. 203; on Selden's and Ussher's support for Walton's Bible, see Toomer, *John Selden*, pp. 800–01.

On Thomas Hyde, see P. J. Marshall's biography in the *Oxford Dictionary of National Biography*. On his career in Oxford, see Henry John Todd, *Memoirs of the Life and Writings of the Right Rev. Brian Walton* (London: F.C. and J. Rivington, 1821), pp. 263–8; 'made great progress' is on p. 267. For Wood's comments on Hyde, see *The Life and Times of Anthony Wood*, vol. II, p. 72, and vol. III, p. 213. Hyde's passion for Oriental languages is noted in Toomer, *Eastern Wisedome and Learning*, pp. 248–9.

'The Chinese language has no analogy': 'Préface des mémoires de la Chine', *Lettres édifiantes et curieuses, écrites des missions étrangères* (Toulouse: Noel-Etienne Sens & Auguste Gaudé, 1810), vol. XVI, pp. 22–4. For Hong Kong Governor George Bonham's contempt for the study of Chinese, see Jack Gerson, *Horatio Nelson Lay and Sino-British Relations, 1854–64* (Cambridge, MA: East Asian Research Center, Harvard University, 1972), p. 31.

Hyde's Chinese manuscript notes are preserved in the British Library in Sloane 853a. The additional Michael Shen manuscripts are in Sloane 4090. I am grateful to Frances Wood for providing me with copies of these files.

Hyde's notes on Chinese chess: Thomas Hyde, *Historia Nerdiludii* (Oxford: At the Sheldonian Theatre, 1694), pp. 195–202; *Mandragorias, seu Historia Shahiludii* (Oxford: At the Sheldonian Theatre, 1694), pp. 159–78, also the inserts following p. 70. These two works together constitute his *De Ludis Orientalium* ('On Oriental Games'). To the extent that Hyde has a reputation today, it is for having written the first European account of Asian board games. His letter on the lack of student interest in his lectures is quoted in Toomer, *Eastern Wisedome and Learning*, p. 298.

'Michael Shun Fo-Çung (for that was his name)': Toomer, *Eastern Wisedome and Learning*, p. 298. 'My Chinese friend Dr Michael Shin-Fo Çungh': Thomas Hyde, *Historia Religionis Veterum Persarum* (Oxford: At the Sheldonian Theatre, 1700), p. 522. Selden's analogy of Oriental languages to the telescope: Toomer, *Eastern Wisedome and Learning*, p. 69.

'Baroque Orientalism': Nicholas Dew, *Orientalism in Louis XIV's France*, p. 6.

'Pseudo-Orientalism': Claire Gallien, *Orient anglais: connaissances et fictions au XVIIIe siècle* (Oxford: Voltaire Foundation, 2011), p. 176.

Hyde's sale of his Oriental manuscripts: Macray, *Annals of the Bodleian Library*, p. 113. His retirement letter to the Archbishop of Canterbury is quoted from Toomer, *Eastern Wisedome and Learning*, p. 298.

4. John Saris and the China Captain
The principal sources for this chapter are the journals of two participants: *The Voyage of Captain John Saris*, ed. Ernest Satow (London: Hakluyt Society, 1900), and *The Diary of Richard Cocks*, ed. Edward Maunde Thompson, 2 vols (London: The Hakluyt Society, 1883; repr. Cambridge: Cambridge University Press, 2010). The source of other EIC journals, letters and reports is *The English Factory in Japan, 1613–1623*, ed. Antony Farrington, 2 vols (London: British Library, 1991). On Thomas Smythe's handling of Saris's erotica, see Timon Screech, '"Pictures (The Most Part Bawdy)": The Anglo-Japanese Painting Trade in the Early 1600s', *The Art Bulletin* 87:1 (March 2005), pp. 50–72.

On Li Dan, see Seiichi Iwao, 'Li Tan, Chief of the Chinese Residents at Hirado, Japan, in the Last Days of the Ming Dynasty', *Memoirs of the Research Department of the Toyo Bunko* 17 (1958), pp. 27–85. I see no reason for supposing, as Robert Batchelor does, that Li Dan had something to do with the Selden map, but then I seem to disagree with Robert about a great deal. His essay 'The Selden Map Rediscovered: A Chinese Map of East Asian Shipping Routes, c.1619', *Imago mundi* 65:1 (2013), pp. 37–63, does the service of introducing the map to cartographic historians but offers hypotheses that may not stand the test of time.

For Martina van Ittersum's blunt assessment of de Groot as a spokesman for Dutch imperialism, see her *Profit and Principle*, pp. 451, 486, 489. Selden's involvement with the Virginia Company is mentioned in Toomer, *John Selden*, p. 320; his purchase of Purchas's manuscripts, pp. 19–20.

Matinga's story is told from the references in Cocks's *Diary*, vol. II, pp. 30, 31, 93, 95, 102, 109, 131, 144–5, 190, 252, 318. Alison Games has also noted the Matinga–Cocks relationship in *The Web of Empire: English Cosmopolitans in an Age of Expansion, 1560–1660* (Oxford: Oxford University Press, 2008), p. 106.

Bankruptcy of the China Captain of Batavia: Leonard Blussé, 'Testament to a Towkay: Jan Con, Batavia and the Dutch China Trade', in *All of One Company: The VOC in Biographical Perspective*, ed. Robert Ross and George Winius (Utrecht: HES, 1986), pp. 3–41.

On Li Dan and Zheng Zhilong, see Tonio Andrade, *Lost Colony: The Untold Story of China's First Great Victory over the West* (Princeton: Princeton University Press, 2011), pp. 22–8.

5. *The Compass Rose*

On Francisco Gomes, see *The Voyage of Captain John Saris*, pp. 50–57.

On the history of the compass in China, see Joseph Needham with Wang Ling, *Science and Civilisation in China*, vol. IV, pt 1 (Cambridge: Cambridge University Press, 1962), pp. 279–334. The account is not entirely satisfactory, as Needham built his explanation on geomantic compasses rather than maritime ones. Had he consulted the Laud rutter to see how marine compass readings actually worked, he would have discovered the logic of trisection.

'Daily necessities come from across the sea': Xiao Ji's preface (1617), in Zhang Xie, *Dong xi yang kao* ('Study of the Eastern and Western Seas') (Beijing: Zhonghua shuju, 1981), front matter, p. 15.

On the Fire Chief, see Needham, *Science and Civilisation in China*, vol. IV, pt 1, p. 292.

William Dampier's description of a Chinese junk: *A New Voyage Round the World* (London: James Knapton, 1699), pp. 412–13.

'The larger ships are over ten metres': Zhang, *Dong xi yang kao*, p. 171.

The practice of pairing compasses fore and aft is mentioned in the classic Ming manual of technical processes published in 1637: Ying-hsing Sung, *Chinese Technology in the Seventeenth Century*, trans. E-tu Zen Sun and Shiou-chuan Sun (University Park: Pennsylvania State University, 1966), p. 177.

'Drop a piece of wood': *Liangzhong haidao zhenjing* ('Two Sea Rutters'), ed. Xiang Da (Beijing: Zhonghua shuju, 1961), p. 6. For a different sailing speed see Cordell Yee, 'Reinterpreting Traditional Chinese Geographical Maps', in *The History of Cartography*, vol. II, bk 2, p. 53, n. 33.

On William Laud, see Charles Carlton, *Archbishop William Laud* (London: Routledge and Kegan Paul, 1987). On Selden's appointment to the Parliamentary committee in 1641 to impeach Laud, see John Aikins, *The Lives of John Selden and Archbishop Usher* (London: Mathews and Leigh, 1812), pp. 113–14.

'My nearer care of J.S. was professed': William Laud, *The Works of William Laud*, ed. James Bliss, vol. III (Oxford: John Henry Parker, 1853), p. 225.

'The Puritan would be judged by the word of God': *The Table Talk of John Selden*, p. 150; see also Barbour, *John Selden*, pp. 255–6.

'A great deale of Learning': Toomer, *Eastern Wisedome and Learning*, p. 108, quoting a letter Laud wrote to the Levant Company in 1634 obliging its agents to purchase manuscripts in the Near East.

'If you make an error of a tiny fraction': Xiang Da, *Liangzhong haidao zhenjing*, p. 21, translated in Needham, *Science and Civilisation in China*, vol. IV, pt 1, p. 287.

Shen Gua on As the Bird Flies is quoted in Needham, *Science and Civilisation in China*, vol. III, p. 576. I have altered the translation.

Will Adams's list of Chinese names for the winds: Farrington, *The English Factory in Japan*, p. 1126.

On portolan charts, see Richard Unger, *Ships on Maps: Pictures of Power in Renaissance Europe* (London: Palgrave Macmillan, 2010), ch. 3. Unger notes that compass roses make their appearance on European charts only in the second half of the fourteenth century (p. 43).

'Very well skilled in Maritim affairs': Selden, *Of the Dominion*, pp. 366–9. My comments are based on the maps in the English edition; these are basically identical to the maps in the original Latin edition. Those in European editions are sometimes altered.

'Ever at home, yet have all countries seen': Ben Jonson, 'An Epistle to Master John Selden', in *The Poetical Works of Ben Jonson*, p. 166.

Stephen Davies: 'The Construction of the Selden Map: Some Conjectures', *Imago mundi* 65:1 (2013), pp. 97–105. This essay is properly titled, as many of its findings are entirely conjectural and more than a few, I fear, mistaken.

6. Sailing from China

'Once you are through the harbour entrance': Zhang Xie, *Dong xi yang kao*, p. 171. Readers interested in reading more on the maritime history of East Asia in this period might begin with my book *The Troubled Empire: China in the Yuan and Ming Dynasties* (Cambridge, MA: Harvard University Press, 2010), ch. 9.

Zhang Xie's depiction of mariners going to sea appears in his *Dong xi yang kao*, p. 170; I have consulted the partial translation of this passage in Needham, *Science and Civilisation in China*, vol. IV, pt 1, pp. 291–2.

'Embark through Five Tiger Gate': Xiang Da, *Liangzhong haidao zhenjing*, p. 49.

The Yuan map showing the sea route to the Indian Ocean starting in Quanzhou is Qingjun's *Broad-Wheel Map of the Frontier Regions*; it is discussed in my *Troubled Empire*, p. 220, and reproduced in Jerry Brotton, *A History of the World in Twelve Maps*), p. 136.

For Japanese place-names in the Laud rutter, see Xiang, *Liangzhong haidao zhenjing*, pp. 97, 99.

'Steered away South-west': *The Voyage of Captain John Saris*, p. 184.

Cocks's commission to Wickham: Farrington, *The English Factory in Japan,* vol. I, pp. 230–34. Additional information on the EIC voyages has been taken from vol. II, pp. 1048–1103, 1586–8; also *Diary of Richard Cocks,* vol. I, pp. 7, 70, 79, 269, 300, 340–48; vol. II, pp. 1–4, 12, 18, 36, 58, 270.

Zhang Xie's account of Wanlaogao: *Dong xi yang kao,* pp. 101–2; of Old Langleishi Bilixi, pp. 89–90.

7. Heaven is Round, Earth is Square

The writing of Coleridge's 'Kubla Khan' is told by John Livingston Lowes in his classic *The Road to Xanadu: A Study in the Ways of the Imagination* (London: Constable, 1930), ch. 19. Coleridge dates the poem to 1797, but Lowes corrects this to 1798. I am grateful to Dorothy Cutting for presenting me with a copy of what has to be the most exuberant study in English literature.

On Samuel Purchas (baptised 1577, d. 1626), see David Armitage's biography in the *Oxford Dictionary of National Biography.* Armitage calls *Purchas his Pilgrimes* 'the bulkiest anti-Catholic tract of the age'. He mentions Purchas's membership in the Virginia Company from 1622 to 1624; Selden's membership is noted in Toomer, *John Selden,* p. 320.

Pantoja: Samuel Purchas, *Purchas his Pilgrimage,* vol. III, pp. 360–61; the Saris map: pp. 401–4.

On the Ming as a 'unified' realm, see Brook, *The Troubled Empire,* pp. 26–9.

John Speed's shire maps (though not identified as such) in Hirado: *The English Factory in Japan,* ed. Farrington, vol. II, p. 1363. Speed remains little studied, but see Ashley Bayton-Williams, 'John Speed: The First Part of an Extensive Biography', online @ http://www.map-forum.com/02/speed.htm, accessed 6 May 2012; see also Martha Driver, 'Mapping Chaucer: John Speed and the Later Portraits', *The Chaucer Review* 36:3 (2002), 228–49: 'very rare and ingenious capacitie' and 'setting this hand free' are quoted on p. 230.

Luo Hongxian's atlas *Guang yutu* ('Enlarged Territorial Atlas') was published in 1555. On Luo's impact on Ming cartography, see Timothy Brook, *The Chinese State in Ming Society* (London: Routledge Curzon, 2005), pp. 47–57. The grid system is examined in Joseph Needham with

Wang Ling, *Science and Civilisation in China,* vol. III (Cambridge: Cambridge University Press, 1959), pp. 539–56.

'He hung a lamp above his desk': Wan Shenglie, 1613 preface to *Tushu bian* ('Documentarium'), 1b.

'What I have striven for is to make present knowledge reliable': Zhang Huang, preface to *Nanchang fuzhi* ('Gazetteer of Nanchang Prefecture') (1588), 1a.

Zhang Huang's maps appear in *Tushu bian* ('Documentarium') (1613), 29.33b–34a, 36b–37a.

On early modern European maps and the problem of the Iberian division of the globe, see Jerry Brotton, *Trading Territories: Mapping the Early Modern World* (London: Reaktion, 1997), ch. 4. Particularly helpful on the Mercator projection is Mark Monmonier, *Rhumb Lines and Map Wars: A Social History of the Mercator Projection* (Chicago: University of Chicago Press, 2004).

'A map of the world that does not include Utopia is not worth even glancing at': Oscar Wilde, 'The Soul of Man under Socialism' (1891).

8. Secrets of the Selden Map

'New found to be an earth inhabited': Ben Jonson: *Masques and Entertainments*, ed. Henry Morley (London: Routledge, 1890), p. 245.

On the use of the twenty-eight mansions as a universal clock, Zhang Huang, *Tushu bian*, 17.69a. Zhang Huang doesn't include the sun and moon in his astral correspondence map, 29.3a–3b, but does insert them in the upper corners of a chart showing how the natural and human realms interact to create political order; ibid., 18.2b–3a. On Yu Xiangdou, see Timothy Brook, *The Confusions of Pleasure: Commerce and Culture in Ming China* (Berkeley: University of California Press, 1998), pp. 163–7, 213–4.

The Works of Heaven and the Inception of Things: Ying-hsing Sung, *Chinese Technology in the Seventeenth Century*, p. 178.

Cordell Yee on scale mapping: 'Reinterpreting Traditional Chinese Geographical Maps'; on 'intuitive sense of underlying form' see his 'Chinese Cartography among the Arts', *The History of Cartography*, vol. II, bk 2, pp. 63–4, 148.

'Coming through Plum Pass as the Evening Clears': I am grateful to Liam Brockey for bringing Ou Daren's poem to my attention.

'The Chineses had refused to trade with the English': *The Voyage of Captain John Saris*, p. 131; Cocks on Chinese plunderers blaming the Dutch and English: *The Diary of Richard Cocks*, vol. II, p. 321.

Imperial edict of 1614 on 'anti-Japanese defence boats': *Qiongzhou fuzhi* ('Gazetteer of Qiongzhou Prefecture') (1619), 8.1b. Qiongzhou Prefecture is today's Hainan Island.

Epilogue

Joan Blaeu's c.1640 portolan chart of the China seas is reproduced in Gnther Schilder and Hans Kok, *Sailing for the East: History and Catalogue of Manuscript Charts on Vellum of the Dutch East India Company* (Houten: HES & De Graaf, 2010), p. 675.

Edmond Halley's dismissal: R. T. Gunther, *Early Science in Oxford* (Oxford: Hazell, Watson and Viney, 1925), vol. III, p. 264.

Curiosities donated to the Bodleian: Macray, *Annals of the Bodleian Library*, pp. 90, 104–5, 107. For human remains, see John Pointer, *Oxoniensis Academia* (London: S. Birt, 1749), p. 157. The Bodleian was not alone in collecting such items. The Royal Society in London possessed, among other things, 'the entire Skin of a Moor, tanned with the Hair on, every part remaining' (noted from the Society's catalogue in Hutton, *A New View of London*, vol. II, p. 666).

Handbill advertising Giolo: 'Prince *Giolo* Son of the King of *Moangis* or Gilolo: Lying under the *Æquator* in the *Long.* of 152 *Deg. 30 Min.* a fruitful Island abounding with rich Spices and other valuable Commodities' (undated, 1692); book about Giolo attributed to Thomas Hyde: *An Account of the Famous Prince Giolo, Son of the King of Gilolo, Now in England: With a Account of his Life, Parentage, and his Strange and Wonderful Adventures* (London: R. Taylor, 1692). Dampier's version of Giolo's story appears in his *A New Voyage*, p. 549. On Giolo in England, see Geraldine Barnes, 'Curiosity, Wonder, and William Dampier's Painted Prince', *Journal for Early Modern Cultural Studies* 6:1 (2006), pp. 31–50.

Arbitration of Dutch and American claims to Miangas: 'Island of Palmas Case' (4 April 1928), *Reports of International Arbitral Awards*, vol. II, pp. 829–71 (repr., New York, 2006).

'The Glory of our Nation for Orientall learning': 'Selden Correspondence', Bodleian Library, Selden supra 108, p. 141, n. 539.

Selden's executors were Edward Heyward, Rowland Jewkes, John Vaughan and Matthew Hale. All but Hale were members of the Inner Temple; see William Cooke, *Students Admitted to the Inner Temple, 1571–1625* (London: F. Cartwright, 1868), pp. 95, 98, 151, 153.

'A handsome white marble Mon': Hutton, *A New View of London*, vol. II, p. 570.

'It were an Ignorance beyond Barbarism': letter from James Howell to John Selden, Selden Correspondence, Bodleian Library, Selden supra 108, p. 218.

Index

Page references for footnotes are followed by f

A
Acapulco 122
Adams, Will x, 77, 103, 120, 131, 143
  Northern Sea route 116, 117–19
Aden 127, 128
Alexander VII, Pope 36, 51
America *see* United States
Anatomy School 177, 180–1
Andace, Captain *see* Li Dan
Arabic 40, 41
Ashmolean Museum 99, 177
Asia North Lambert Conformal Conic Projection 159, 187
Asia with the Islands Adjoining Described (Speed) 159, 186
astrolabes 98–9
Atlas Sinensis 51

B
Baker, Margaret 25
Bantam xii, 29, 72, 169–70, 171
Barlow, Thomas 19, 20, 21, 54
Bartholomew Fayre (Jonson) 41–2
Batavia xii, 84f, 120, 169, 170
Batchelor, Robert 11
Beaumont, Francis 23
Besero, Don Fernando 74

Best, Thomas 171
Bible
  creation of the world 153
  King James Bible 135
  and maps 135, 144
  Polyglot 55, 64
Blaeu, Joan 51, 176
Blaeu, Willem 51
Blocq, Adriaan Martens 73, 76
Blussé, Leonard 84–5f
Bodleian Library 11
  astrolabes 98–9
  catalogue 54, 58, 64
  collecting practices 54–5
  Confucius Sinarium Philosophus 48–9, 52
  Hyde 54, 55, 56
  James I 51
  James II 46–9
  Keepers 64
  Laud rutter 97, 98
  Polyglot Bibles 55
  Selden End 20, 46, 47
  Selden map xx, xxii, 10–11, 19, 21–2, 97, 176–7
  Selden's library 19–22
  Shen 58
  Wood 54
Bodley, Thomas 11
Bonham, George 57
Book of Changes 104

Borneo 12, 161
boxing the compass 103–5, 183–5
Boyle, Robert 63
Boym, Michael 49
Brazil 4, 13
Brunei 4, 122
Buckingham, duke of 27–8, 32,
   33–4, 35

C
Calicut 127–8
calipers 108–9
Camden, William 135
Celebes 89
Charles I 35
   and *The Closed Sea* 36, 38, 98
   King's Evil 46
   and Laud 98
   ship money 36, 99–100
Charles II 46
China
   books 11
   compasses 88–9, 100–5, 183–5
   East India Company 90
   Eastern Sea route 126
   Friendship Pass 1–2
   Hainan Island incident 6–10
   James II 47
   and Japan 77.78
   Jesuits 48, 49, 51–3, 57
   maps 2–6, 12, 18, 87–8, 105, 129,
      131–4, 138, 140–3, 159–60, 169
   mariners 14, 18
   measurements 88
   religion 48–9, 94
   Ryukyu Islands 118
   sailing routes 110–14

Saris 168
seafaring 90–7
Selden map xx–xxii, 10–15,
   155–9
Senkaku Islands 114
South China Sea xxii, 3–4,
   6–10, 13, 14
   and Taiwan 86
   Ternate 125
   trade 78–81, 84–5
   Western Sea route 126–8
China Captain 77, 84f
Chinese Convert, The (Kneller)
   52–3
Chinese language 57–62, 65–6,
   105f
Clavell, James 81
Closed Sea, The (Selden) 32–3,
   36–40, 98, 99, 106–8
Clove 67, 68, 73, 74, 116
cloves 72–3, 75
Cocks, Richard x, 119–20, 126, 131
   China 90
   and Dutch 73
   Hyogo 115
   and Li brothers 78, 79–84
   Northern Sea route 116, 117, 119
   oil paintings 71–2
   quadrant 88
Codex Mendoza 171
Coleridge, Samuel Taylor 64,
   129–30, 134, 147, 152, 157
Columbus, Christopher xx
Comedy of Errors (Shakespeare)
   23
'Coming through Plum Pass as
   the Evening Clears' (Ou) 167–8

compass roses 105–6, 163–4
  Selden map 108, 109, 163–4
compasses 88–9, 95–7
  China 100–5, 183–5
  Chinese vs Mediterranean
    103–4
  Laud rutter 97, 100–3
  Selden 101–2
*Complete Course for a Myriad
  Practical Uses* (Yu) 157
Confucius Sinarium Philosophus
  48, 51–2, 53
conic projection 159, 187
Cotton, Robert 135
Couplet, Philippe 49–52, 53, 63

D
Damariñas, Gómez Pérez 124, 125
Damariñas, Luis Pérez 124
Dampier, William 93–4, 179
Davies, Stephen 109
de Groot, Huig 14, 28, 30, 31,
  75–6, 131, 154
  Free Sea, The 28, 30, 32, 38, 39,
    72, 86
  and Selden 32, 35, 36, 38
Denmark 32
Deuteronomy 42
Dew, Nicholas 64
Diaoyu Islands xxii, 114
Dittis, Andrea *see* Li Dan
divine right of kings 25–6
Djofar 127, 128
Documentarium (Zhang Huang)
  139–42, 146, 156–7
Dong xi yang kao (Zhang Xie)
  92–7, 110–14, 123–5, 128

Dongyang *see* Eastern Sea
Donne, John xxiii, 23
Downton, Nicholas 171
Dürer, Albrecht xix
Dutch *see* Netherlands
Dutch East India Company 28–
  30, 31, 38, 39, 86
  China 84, 84–5f
  portolan chart 176
  Spice Islands 72–6
  Ternate 124–6

E
East China Sea xxii, 114
East India Company 68
  China 79–81, 90
  Cocks 81–2
  commanders 170–1
  Eastern Sea route 126
  Japan 69–72, 77–83, 84, 143
  Northern Sea route 116–20
  pilots 89–90
  pornography scandal 69–70
  Saris 67–9, 72–6, 77
  shop 68
  Spice Islands 72–6
  women 83–4
Eastern Sea route 114, 121–6, 172
Elizabeth I 30, 68, 135
Enlarged Territorial Atlas (Luo)
  137–8, 140

F
Field Division (*fenye*) 156–7
Fire Chief 93–4
  see also pilots
Fischer, Joseph xix, xx

Franks 123–4
Free Sea, The (de Groot) 28, 30,
    32, 38, 39, 72, 86
Friendship Pass 1–2

G
Galileo 51, 64
Gallien, Claire 64
General Map of Chinese and
    Barbarians within the Four
    Seas (Zhang Huang) 140
General Topographical Map by
    Province (Yu) 157–8
geo-referencing 166
Gift of God 118
Giolo x, 177–81
GIS (Geographic Information
    System) 166
globes 47, 143
Gobi Desert xxi
Gomes, Francisco 74, 75, 88
Goto Archipelago 114
gourds 147
Greville, Fulke 135
Grotius *see* de Groot, Huig
Guang yutu (Luo) *see Enlarged*
    Territorial Atlas
Gui Island 92, 111
Gypsies Metamorphosed, The
    (Jonson) 33–4

H
Hafner, Hermann xx
Haicheng 92
    see also Moon Harbour
Hainan Island xii, 16, 120
    on the Selden map 157

Hainan Island incident 6–10, 17
Hakluyt, Richard 132, 152, 170,
    171, 176
Halley, Edmond 176
Hebrew 40–1
Helliwell, David 10–11, 61, 182
herring fishery 30–2, 36
Hippon, Anthony 170–1
Hirado xii, 71–2, 77–8, 79, 82–3,
    89, 116
Historie of Tithes, The (Selden)
    25–7, 32, 41, 99
History of Science Museum,
    Oxford 99, 101–2
History of the Religion of the Old
    Persians (*Historia Religionis*
    Veterum Persarum) (Hyde) 54,
    55, 56, 63–4, 105f
Hoàng Sa Islands xxii, 16–17, 167
Hondius, Jodocus 51, 131–2, 135,
    142
Hondius his Map of China 131–2,
    135, 142, 146
Hormuz 127, 128
House of Commons 33, 35
House of Lords 33
Howell, James 43
Huang Ming yitong fangyu beilan
    132–4
Hyde, Thomas x, 18, 48–9, 52,
    53, 54, 63
    Bodleian Library 64, 66, 177
    compass rose 105
    and Giolo 179–80
    History of the Religion of the
        Old Persians 54, 55, 56, 63–4,
        105f

Oriental languages 54, 55, 56–7,
 64, 65–6
Polyglot Bible 55–6
portrait 64–6, 128
Selden map 59–61, 126, 156,
 159, 170, 173
and Shen 54, 57–63
Hyogo 115

I
I Ching 104
Indonesia 12, 179f
Inner Temple 25, 181
innocent passage 8, 9, 13
international law 13–14,
 28–30
see also law of the sea
Iwao, Seiichi 78

J
Jakarta *see* Batavia
James I 25–6, 35, 51
 Bible 135
 China 80–1
 and de Groot 30, 31
 East India Company 68, 70
 and Jonson 23, 24, 33–4, 149,
  169
 King's Chambers 107, 169
 and Laud 98
 and Raleigh 40
 and Selden 22, 23, 25, 26–7, 32,
  33, 34–5, 169
James II 46–8
 *Confucius Sinarium
  Philosophus* 48–9, 52
 and Shen 48, 52

Japan
 Chinese community 78
 East India Company 68, 69–72,
  77–83, 84, 143
 Li Dan 77, 78, 86
 pilots 89
 Ryukyu Islands 118
 Selden map 12, 115–16, 169
 Senkaku Islands 114
Java 169–70
Jesuits 49, 50, 51–2, 57
Jewkes, Rowland 181–2
Johor 29, 30, 43, 120, 126
Jonson, Ben x, xxiii, 23–5, 33–4
 Bartholomew Fayre 41–2
 and East India Company 68
 and James I 23, 24, 33–4, 149,
  169
 *News from the New World
  Discovered in the Moon* 149–
  50, 151–2
 and Raleigh 40–1f
 and Selden 23–4, 42, 108–9
junks 77, 93–4

K
Kagoshima 115
Keeling, William 171
Khubilai Khan 147
King's Chambers 106–8, 169
King's Evil 45–6
Kneller, Godfrey 52–3
Kobe 115
Kremer, Gerard *see* Mercator,
 Gerard
Kubla Khan (Coleridge) 129–30,
 134, 147, 157

Kuroshio Current 114–15

L
Lancaster, James 171
Laud, Archbishop William x, 36, 49, 97–100
Laud rutter 97, 100–3, 158, 163
  sea routes 111, 112, 128
law of the sea 8–9, 13–15, 38
  de Groot 28–30
  Selden 32–3, 36–9
Lecomte, Louis 14
Lee, Martha 164–5, 166
Li Dan x, 77–8, 79–81, 83, 84–5, 86, 112, 116
Li Huayu 78, 79–81, 112
liberty 15
Library of Congress xix, xxi–xxii
Lindisfarne 107
longitude 146
Longzishaji *see* Nagasaki
Louis XIV 46, 50–1
lunar mansions 157
Luo Hongxian 51, 136–8, 140, 142
Luzon 169

M
mace 72
Magdalen College 46
Magna Carta 34
magnetic pole 164–5
Malacca 29, 121, 165
Manchus 134
Manila 78, 121, 169
maps 5
  of China 2–6, 12, 18, 105, 129, 131–4, 136, 137–8, 140–3

in *The Closed Sea* 106–8
Hondius 131–2, 135
Luo's atlas 137–8, 142
Mercator 144–6
Ortelius 146
portolan charts 106
projection 143–6
Saris map 132–4, 138, 142, 146–7, 170
Speed 134–6, 162
Waldseemüller xix–xx, xxi–xxii
Yu 157–8
Zhang 157
see also Selden map
Mare Clausum (Selden) 32–3, 36–40, 98, 99, 106–8
Mare Liberum (de Groot) 28, 30, 32, 38, 39, 72, 86
Marin, Damien 117
maritime law *see* law of the sea
Matinga 82–3
Matyan 72, 73
Mencius 51
Mengxi bitan (Shen Gua) 103
Mercator, Gerard 144–6
Mercator projection 145–6, 165
Miangis 179
Middleton, David 171
Middleton, Henry 171
Milton, John 43
Ming dynasty 16, 18, 134
  maps 11–12
  sun and moon symbolism 159
  trade policy 167–9
Minte, Robert 161
Mission PR32 6–10
Moluccas *see* Spice Islands

Mongols 134
moon 158–9
Moon Harbour 92, 111, 112–13, 121, 139
Muscovy Company 69–70

N
Nagasaki 78, 115–16, 169
Nansha xxii, 4, 167
Nanyang *see* South China Sea
navigation
    night sky 158
    see also compasses
Nedham, Marchamont 38f
Needham, Joseph 89, 93, 163
Netherlands
    herring fishery 30–2, 36
    Jakarta 170
    and Japan 78
    Miangis 179
    Selden map 60–1
    Spice Islands 72–6, 168
    Ternate 124–6, 173
    trade 28–31
News from the New World Discovered in the Moon (Jonson) 149–50, 151–2
North by Northwest 105
North Sea 32, 36
Northern Sea route 114–20, 172
nutmeg 72

O
Okinawa xii, 114, 118
On the Law of Prize or Booty (de Groot) 30

Oriental studies 40–1, 42–3, 54–8, 66, 97, 100, 153
    rise and fall in England 63–4
Ortelius, Abraham 131, 142, 146
Osborn, Lieutenant Shane 6, 7, 8
Ou Daren 167–8
Oxford University *see* University of Oxford

P
Pantoja, Diego de 131, 133, 138, 141
Paracel Islands xii, xxii, 16–17, 167
Pen Conversations from Dream Brook (Shen Gua) 103
Pepys, Samuel 38f, 181
Persian 41
Philippines 4, 121, 169
    Selden map 12, 161
pilots 89–90, 93–6
Pinochet, Augusto 33
platts/plotts *see* sea charts
Polyglot Bibles 55–6, 64
portolan charts 106, 107–8, 144, 176
portolanos see rutters
Portugal
    and Dutch 28–30
    East Indies 28–30, 39
    and Japan 78, 115–16
    mariners 14
    Paracel Islands 4
    seafaring 96
    Spice Islands 74
    Treaty of Tordesillas 13, 146
Poynter, Theophilus 180

Pratas Reef 166–7
Pring, Martin 171
projection 143–6, 159
Prospect of the Most Famous Parts of the World (Speed) 131, 134–6
Purchas, Samuel x, 139, 141, 142–3, 152, 176, 129–31
  Hondius his Map of China 131–2, 135, 146
  Saris map 132–4, 138, 142, 146–7, 170
  and Selden 130–1
Purchas his Pilgrimage 129–31
Purchas his Pilgrimes 130, 131–4, 139, 146–7, 170

Q
qi 156–7
Qing dynasty 134
Quanzhou 112

R
Raleigh, Walter 40, 40–1f
Raleigh, Wat 41f
Rangoon 127
rhumb lines 106, 107–8, 109, 144, 145
Ricci, Matteo 133, 141, 146
roses *see* compass roses
route guides *see* rutters
rubber-sheeting 166
Rubens, Peter Paul 35
rulers 108, 162, 163
Ruscelli, Girolamo 141
rutters 96–7
  Laud rutter 97, 100–3
Ryukyu Islands xii, 114, 118

S
Said Barakat 125
sailing speed 65–6, 162–3
Santa Catarina 29–30, 43, 120
Saris, John xi, 67–9, 70–1, 78, 81, 84, 141, 151–2
  China trade 168
  compasses 88
  Eastern Sea route 126
  Japan 77
  map 132–4, 138, 142, 146–7, 170
  Northern Sea route 116, 117
  and pilots 89
  and Purchas 131
  sea charts 96
  and Selden map 171–3
  Spice Islands 72–6
Satsuma 115
Sayers, Edmund 96, 117, 118–19, 120
Schöner, Johannes xx, xxi, xxiv
scrofula 45–6
Sea Adventure 117–19
sea routes 161–2, 163–6
Selden, John xxiii, 22–3, 25, 64, 151–2
  arrests 33–7, 98
  and Buckingham 27–8, 35
  The Closed Sea 32–3, 36–40, 98, 99, 106–8
  compass 101–2
  and de Groot 32, 38
  The Historie of Tithes 25–7, 32, 41, 99
  and Jonson 23–4, 42
  and Laud 36, 98–9
  law of the sea 14–15

liberty 15
library 19–22
Oriental languages 40–1, 42–4, 55
poetry xxiii, 23
politics 33, 34–5, 98, 99–100
Polyglot Bible 55
and Purchas 130–1
royal audience 22, 23, 25–7, 169
Titles of Honor 153–4
tomb 181–2
will 20–1, 169
Selden, John (the Minstrel) (father) 25
Selden map xx–xxiv, 19, 39, 153–5, 175–6, 182
    accuracy 11–18, 159–63, 186–7
    in Bodleian Library 10–11, 19, 20–1, 22, 176–7
    China 155–9
    compass rose 87–8, 108, 109
    date 173
    Johor 43
    magnetic signature 163–6
    origin 169–74
    ruler 88, 108–9
    sea routes 111, 113, 114–28
    Shen and Hyde 59–61
    South China Sea 166–9
    Xanadu 146–7
Selden types 40
Senkaku Islands 114
Shakespeare, William xxii–xxiii, 150–2
shapeshifting foreigners 122
Shapleigh, Alexander 171

Shen Fuzong, Michael xi, 48, 49–54, 57–63, 65, 66, 159
    compass rose 105f
Shen Gua 103
ship money 36, 100–1
Shogun (Clavell) 81
Siam 80, 117
Silva, Jerónimo de 74, 75
Singapore Strait 29, 39, 43, 120–1
Sloane, Hans 53
Smythe, Thomas 67, 68–70, 71, 72
Song Yingxing 158f
South China Sea (Southern Sea) 3–4, 9
    Hainan Island incident 6–10
    Paracel Islands xxii, 4, 16–17
    Selden map 12–13, 14, 16–17, 166–9, 172
    Spratly Islands xxii, 4
sovereignty claims 14, 39, 164, 167, 169
    British 32, 36, 107
    Chinese xxi, 5, 6, 9
Spain
    and Philippines 78
    Spice Islands 72, 74–5
    Ternate 124–6, 173
    Treaty of Tordesillas 13, 146
Speed, John xi, 134–5, 152
    Asia with the Islands Adjoining Described 159, 162, 186
    Prospect of the Most Famous Parts of the World 131, 134–6
speed log 96
Spice Islands 29, 61, 72–6
    see also Ternate

spice trade 12, 28–30, 31, 72–6, 97
Spratly, Richard 4
Spratly Islands xii, xxii, 4, 167
stars 158
Stiglitz, Marinita 161
Study of the Eastern and Western Seas (Zhang Xie) 92–7, 110–14, 123–5, 128
Sugiyama, Keisuke 161
Sulu Archipelago 123
Sumatra 12, 161
sun 158–9
Sunda *see* Bantam
Syriac 41

T
Taiwan 85–6, 121
Taiwan Strait 113
Talbot, Elizabeth 22–3, 181
tattoos 177–8, 180–1
Tempest, The (Shakespeare) xxii–xxiii, 150–2
Temple xii, 25, 181
Ternate xiii, 72, 74–5, 123–6, 173
terra nullius 9, 13, 39, 151, 154
territorial water 8–9, 106–7
Thevet, André 171
Tiangong kaiwu (Song) 158f
Tidore 72, 74–5, 88
tithes 25–7
Titles of Honor (Selden) 153–4
Tokugawa Ieyasu 70–1, 77
Toomer, Gerald 37–8, 56
trade 86
　with China 78–81, 168
　Northern Sea route 116–20

see also Dutch East India Company; East India Company
Treaty of Tordesillas 13–14, 146
True Law of Free Monarchies, The (James I) 26
Turkish 41
Tushu bian (Zhang Huang) *see Documentarium*
Twelfth Night (Shakespeare) 23

U
Unified Terrestrial Realm of the Ming Empire Complete at a Glance 132–4
United States
　Hainan Island incident 6–10
　Miangis 179
　on Waldseemüller map xix–xx, xxi–xxii
University of Oxford
　Anatomy School 177
　Ashmolean Museum 99, 177
　History of Science Museum 99, 101
　Laud 99, 100
　Magdalen College 46
　see also Bodleian Library
Ussher, Archbishop James 40, 41, 55, 152–3
utopia 147

V
van Heemskerck, Jacob 29–30, 43, 75, 120, 125
van Ittersum, Martine 73, 76
Verde Island Passage 121–2

Vespucci, Amerigo xx
Vietnam
    Friendship Pass 1–2
    Paracel Islands 4
    Selden map 12, 161, 162
    South China Sea 4
viewing the land from the sea 162, 168
Villiers, George *see* Buckingham, duke of
Virginia Company 69–70, 131
VOC *see* Dutch East India Company

W
Waldseemüller, Martin xix–xx, xxi–xxii
Walton, Brian 55, 56, 64
Wang Wei, Lieutenant Commander 6–10
Wanlaogao *see* Ternate
Wanli, Emperor 79, 138, 139
Wanyong Zhengzong (Yu) 157
Western Sea route 114, 120–1, 122, 126–8, 172
Whaw, Captain *see* Li Huayu
Wickham, Richard 117, 118
Wilde, Oscar 147–8
Wood, Anthony 19, 20, 47, 48, 54, 180
Wood, Frances 62
Works of Heaven and the Inception of Things (Song) 158f

X
Xanadu 129–30, 147–8, 157
Xiamen 111
Xiang Da 95, 162–3
Xisha xxii, 16–17, 167
Xu, Candida 53
Xu, Paolo 53

Y
Yee, Cordell 10, 159–60
Yellow River 156
Yu Xiangdou 157–8
Yunnan 161

Z
Zhang Huang xi, 138–9, 143, 146
    Documentarium 139–42, 146, 156–7
Zhang Tingbang 91–2
Zhang Xie xi, 91–7, 110–14, 128, 173
    Eastern Sea route 121, 123–5
Zhangzhou 111, 112
    see also Moon Harbour
Zheng Chenggong 85–6
Zheng He 101, 112, 128
Zheng Zhilong 85
zhenjing see rutters
Zhou, Duke of 100
Zhu Siben 137–8